ADVANCED GEOSCIENCE REMOTE SENSING

Advanced Geoscience Remote Sensing

Editor:

Maria Tattaris, Ph.D.

ArclerPress

www.arclerpress.com

Advanced Geoscience Remote Sensing
Editor: Maria Tattaris, Ph.D.

708 3rd Avenue, 6th Floor
New York
NY 10017
United States of America
www.arclerpress.com

ISBN: 978-1-68094-477-8

Library of Congress Control Number: 2016951164

About the Editor

Maria Tattaris, Ph.D.

Maria Tattaris has a background in Mathematics and a PhD in remote sensing within the area of atmospheric sciences, awarded in 2012. Her interests include field spectroscopy, image processing, spectral analysis, field data collection, and investigating new data analysis methods for remote sensing techniques, all within the areas of environmental, plant, atmospheric and related sciences.

Contents

List of Contributors..xi

Preface..*xvii*

Introduction..*xix*

Chapter 1 Remote Sensing Study of Glacial Change in the Northern Patagonian Icefield..1

1. Introduction..1
2. Study Area..2
3. Data and Methods..4
4. Results..6

Discussion and Conclusions..9

References..10

Chapter 2 Snow Depth Retrieval With Uas Using Photogrammetric Techniques..15

1. Introduction..15
2. Study Area and Data..18
3. Methods..22
4. Results..27
5. Discussion and Conclusions..32

Conclusion..33

Acknowledgments..34

References..34

Chapter 3 Change Detection of Desert Sand Dunes: A Remote Sensing Approach..39

1. Introduction..39
2. Methodology..41
3. Results and Discussion..43

Conclusions..50

References..52

Chapter 4 Analysis of Mountain Rangeland Changes, Using Remotely-Sensed Data..55

1. Introduction..55
2. Study Area..56
3. Methodology..56
4. Results and Discussion..58

Conclusion ... 60
References ... 61

Chapter 5 Continental Collision Structures and Post-Orogenic Geological History of the Kangerlussuaq Area in the Southern Part of the Nagssugtoqidian Orogen, Central West Greenland ... 63
1. Introduction ... 63
2. Background: The Development of the Nagssugtoqidian Orogen and the Study Area ... 64
3. Regional Lineament Mapping ... 68
4. Field Observations From Site Areas A and B ... 69
5. Geological Evolution of the Area ... 73
6. Discussion ... 75
Summary ... 77
Acknowledgments ... 78
References ... 78

Chapter 6 Uav Thermal Infrared Remote Sensing ... 81
1. Introduction ... 81
2. Methods ... 82
Conclusions ... 88
Acknowledgements ... 88
References ... 89

Chapter 7 Integrated Assessment of Forest Cover Change and Above-Ground Carbon Stock in Pugu and Kazimzumbwi Forest Reserves, Tanzania ... 93
1. Introduction ... 93
2. Materials and Methods ... 94
3. Methods ... 94
4. Results and Discussion ... 98
Conclusions ... 102
Acknowledgements ... 104
References ... 104

Chapter 8 Evaluating Biomass of Juniper Trees (Juniperus Pinchotii) From Imagery Derived Canopy Area Using The Support Vector Machine Classifier ... 107
1. Introduction ... 107
2. Materials and Methods ... 109
3. Results ... 113
4. Discussion ... 119
Conclusion ... 122
Acknowledgements ... 122

References 123

Chapter 9 Using Spot and Aerial False-Color Infrared (FCIR) Imagery 129
1. Introduction 129
2. Study Area 130
3. Materials and Methods 131
4. Results 133
5. Discussion and Conclusions 137
Acknowledgments 138
Abbreviations 139
References 139

Chapter 10 Mapping and Monitoring of Dust Storms in Iran by Fuzzy Clustering and Remote Sensing Techniques 141
1. Introduction 141
2. Material and Methods Case Study 142
3. Methodology 144
4. Results and Discussion 146
Conclusion 150
References 152

Chapter 11 Detection of Shoreline Changes on the Visakhapatnam Coast, Andhra Pradesh, from Multi-Temporal Satellite Images 155
1. Introduction 155
2. Study Area 156
3. Material and Methodology 156
4. Results and Discussion 159
5. Summary and Conclusions 161
Acknowledgements 164
References 164

Chapter 12 Impacts of Rainstorms During Austral Winter in Sao Paulo State, Brazil: A Case Study 167
1. Introduction 167
2. Data and Methodology 168
3. Results 169
4. Summary 182
Acknowledgements 183
References 183

Chapter 13 Impacts of Urbanization on Water Resources and Vegetation on the Delta Region of Tamilnadu Using Remote Sensing and GIS 185
1. Introduction 185

2. Study Area Description ... 186
3. Data Sources and Methodology ... 187
4. Results ... 188
Conclusion ... 190
References ... 190

Chapter 14 Landsat Image Analysis of Tree Mortality in the Southern Sierra Nevada Region of California During the 2013-2015 Drought ... 191

1. Introduction ... 191
2. Study Area ... 193
3. Methods ... 193
4. Data Analysis ... 194
5. Results ... 195
6. Discussion ... 199
Conclusions ... 201
Acknowledgements ... 202
References ... 202

Chapter 15 Photogrammetric Methodology for the Production of Geomorphologic Maps: Application to the Veleta Rock Glacier (Sierra Nevada, Granada, Spain) ... 207

1. Introduction ... 207
2. Characteristic of the Veleta Rock Glacier ... 208
3. Geomatic Methods ... 208
4. Computer Vision for Photogrammetric Reconstruction ... 211
5. Results and Discussion ... 214
Conclusions ... 217
Acknowledgements ... 218
References ... 218

Index ... 221

List of Contributors

Lucy Dixon
Department of Geosciences, Mississippi State University, Starkville, MS, USA

Shrinidhi Ambinakudige
Department of Geosciences, Mississippi State University, Starkville, MS, USA

Benjamin Vander Jagt
School of Earth Sciences & Byrd Polar Research Center, The Ohio State University, Columbus, OH 43202, USA

Arko Lucieer
School of Land and Food, University of Tasmania, Hobart, TAS 7001, Australia

Luke Wallace
Math and Geospatial Sciences, RMIT, Melbourne, VIC 3000, Australia

Darren Turner
School of Land and Food, University of Tasmania, Hobart, TAS 7001, Australia

Michael Durand
School of Earth Sciences & Byrd Polar Research Center, The Ohio State University, Columbus, OH 43202, USA

Surender Varma
Center of Studies in Resources Engineering, IIT Bombay, Mumbai, India

Vatsal Shah
Natural Resources Engineering (Range Management), Isfahan University of Technology, Isfahan, Iran

Biplab Banerjee
Center of Studies in Resources Engineering, IIT Bombay, Mumbai, India

Krishna Mohan Buddhiraju
Center of Studies in Resources Engineering, IIT Bombay, Mumbai, India

Ali A. Torahi
Department of Remote Sensing and GIS, Faculty of Geographical Sciences, Kharazmi University, Tehran, Iran

Elham Manuchehri
Natural Resources Engineering (Range Management), Isfahan University of Technology, Isfahan, Iran

Jon Engström
Geological Survey of Finland, P.O. Box 96, SF-02151 Espoo, Finland

Knud Erik S. Klint
Geological Survey of Denmark and Greenland, Øster Voldgade 10, DK-1350 Copenhagen K, Denmark

Stefania Amici
Istituto Nazionale di Geofisica e Vulcanologia, Sezione di Roma, Rome, Italy

Matteo Turci
Interdepartmental Center for Industrial Research in Aerospace (CIRI Aeronautica), University of Bologna, Bologna, Italy

Salvatore Giammanco
Istituto Nazionale di Geofisica e Vulcanologia, Osservatorio Etneo, Catania, Italy

Letizia Spampinato
Istituto Nazionale di Geofisica e Vulcanologia, Osservatorio Etneo, Catania, Italy

Fabrizio Giulietti
Interdepartmental Center for Industrial Research in Aerospace (CIRI Aeronautica), University of Bologna, Bologna, Italy

Japhet J. Kashaigili
Sokoine University of Agriculture, Morogoro, Tanzania

Makarius V. Mdemu
Ardhi University, Dar es Salaam, Tanzania

Augustino R. Nduganda
Tanzania Meteorological Agency, Dar es Salaam, Tanzania

Boniface P. Mbilinyi
Sokoine University of Agriculture, Morogoro, Tanzania

Mustafa Mirik
Texas A&M AgriLife Research (TAMAR), Vernon, USA

Sriroop Chaudhuri
Texas A&M AgriLife Research (TAMAR), Vernon, USA

Brady Surber
Texas A&M AgriLife Research (TAMAR), Vernon, USA

Srinivasulu Ale
Texas A&M AgriLife Research (TAMAR), Vernon, USA

Robert James Ansley
Texas A&M AgriLife Research (TAMAR), Vernon, USA

Alvan Karlin
Southwest Florida Water Management SWFWMD, 2379 Broad Street, Brooksville, FL 34604, USA

Mark Fulkerson
Southwest Florida Water Management SWFWMD, 2379 Broad Street, Brooksville, FL 34604, USA

Gene Altman
Southwest Florida Water Management SWFWMD, 2379 Broad Street, Brooksville, FL 34604, USA

Batool Z
Department of Physical Geography, University of Mohaghegh Ardabili, Ardabil, Iran

Sayyad A
Department of Physical Geography, University of Urmia, Urmia, Iran

Kannan R
Andhra Pradesh Space Applications Centre, Hyderabad, India

Kanungo A
Andhra Pradesh Space Applications Centre, Hyderabad, India

Murthy MVR
Andhra Pradesh Space Applications Centre, Hyderabad, India

Franchito SH
Centro de Previsão de Tempo e Estudos Climáticos, CPTEC, Brazil

Gan MA
Centro de Previsão de Tempo e Estudos Climáticos, CPTEC, Brazil

Rao VB
Centro de Previsão de Tempo e Estudos Climáticos, CPTEC, Brazil

Santo CME
Centro de Previsão de Tempo e Estudos Climáticos, CPTEC, Brazil

Conforte JC
Centro de Previsão de Tempo e Estudos Climáticos, CPTEC, Brazil

Pinto O
Grupo de Eletricidade Atmosférica, ELAT Instituto Nacional de Pesquisas Espaciais, INPE C.P. 515, 12245 970, São José dos Campos, SP, Brazil

Arunpandiyan M
Department of Environmental Management, Bharathidasan University, Tiruchirappalli – 24, India

Aarthi M
Department of Environmental Management, Bharathidasan University, Tiruchirappalli – 24, India

Vidyalakshmi R
Department of Environmental Management, Bharathidasan University, Tiruchirappalli – 24, India

Savaridhos RJ
Department of Environmental Management, Bharathidasan University, Tiruchirappalli – 24, India

Prashanthi Devi
Department of Environmental Management, Bharathidasan University, Tiruchirappalli – 24, India

Christopher S Potter
NASA Ames Research Center, Mail Stop 232-21, Moffett Field, CA, USA

Javier de Matías
Departamento de Ingeniería de Sistemas Informáticos, University of Extremadura, Cáceres, Spain

José Juan de Sanjosé
Departamento de Expresión Gráfica, University of Extremadura, Cáceres, Spain

Gonzalo López-Nicolás
Departamento de Informática e Ingeniería de Sistemas (DIIS) and Instituto de Investigación en Ingeniería de Aragón (I3A), University of Zaragoza, Zaragoza, Spain

Carlos Sagüés
Departamento de Informática e Ingeniería de Sistemas (DIIS) and Instituto de Investigación en Ingeniería de Aragón (I3A), University of Zaragoza, Zaragoza, Spain

José Jesús Guerrero
Departamento de Informática e Ingeniería de Sistemas (DIIS) and Instituto de Investigación en Ingeniería de Aragón (I3A), University of Zaragoza, Zaragoza, Spain

Preface

Nowadays, advanced remote sensing technology plays tremendous roles to build a quantitative and comprehensive understanding of how the Earth system operates. The advanced remote sensing technology is also used widely to monitor and survey the natural disasters and man-made pollution. Our planet is nowadays continuously monitored by powerful remote sensors operating in wide portions of the electromagnetic spectrum. Geoscience, or earth science, covers all areas of sciences that relate to planet Earth. Such fields include soil science, atmospheric science, climate science, geochemistry, geobiology, geology and oceanography, to name just a few. Research into the area of geosciences allows us to better understand the Earth's past, present and future. This is particularly important now as many parts of the planet, and the interactions between them, are changing at a rapid rate, for example oceans, landscapes the atmosphere and climate. Remote sensing is a tool with the potential to improve research within this area due advantages such as the possibility to cover large areas at a high spatial resolution, repeatability of measurements allowing for the monitoring through time, non-intrusiveness of measurements and that data can be obtained/measured quickly compared to manual data collection equivalents.

This book, Advanced Geoscience Remote Sensing, presents research topics that make use of remote sensing to investigate key topics within the area of geosciences, for example glacial changes and mapping, land-cover change, carbon stock estimation, model verification, the mapping and monitoring of dust storms, shoreline changes, evaluating impacts of heavy rainfall, impacts of urbanization on vegetation and water resources, tree mortality rates, snow depth estimates, changes in desert sand dunes and thermal mapping of volcanoes. Remote sensing data types and techniques used for the above studies include multispectral satellite imagery (e.g. Landsat, MODIS, ASTER and SPOT), data captured using Unmanned Aerial Vehicles (UAVs), radar imagery, the generation of DEMs and clustering methods. While rigorous enough for the most experienced of scientists, the techniques are well designed and integrated, making the book's content intuitive, clearly presented, and practical in its implementation. It also identifies that there should be a carefully planned effort on research and application development for the advanced geoscience and related engineering remote sensing techniques. An overview of the most advanced remote sensing systems and algorithms is presented and introduces the most advanced image technologies and their applications for both classification and analysis with up-to-date examples of successful projects and case studies. Chapters are contributed by renowned global scientists and experts in the remote sensing field. Techniques used in geologic mapping and mineral and hydrocarbon exploration, image analysis used in mine development and petroleum exploitation, site evaluation, ground water development, surface water monitoring and geothermal resource exploitation are emphasized. It also exhibits how imagery is used to establish environmental baselines; monitor land, air, and water quality; map hazards; and determine the effects of global warming. This book will be of valuable for students, researchers and practitioners, as well as working professionals in petroleum, mining, groundwater, surface water, engineering, and environmental projects.

Editor

Maria Tattaris, ph.d.

INTRODUCTION

Geoscience, or the study of the Earth, is broken into a broad spectrum of specialized sub disciplines, including geology, mineralogy, volcanology, petrology, geochemistry, geophysics, geomorphology, paleontology, stratigraphy, structural geology, engineering geology, sedimentology, edaphology, and pedology. These classifications include scientific researchers concerned with identification of the physical characteristics of materials such as rocks, minerals, and chemical compositions found on the Earth's surface.

Remote sensing technology is an effective and widely established analytical method for geology and mineral exploration, and has proven extremely beneficial by providing access to dangerous or previously inaccessible sites. Aerial imagery acquired from hyper spectral and multispectral imaging sensors such as Landsat, ASTER, AVIRIS, HyMap and Hyperion is applied to geological surveys, alteration zones mapping, and geomorphology applications. Satellite imagery correlated with ground truth data is also used for assessments of change due to natural events such as volcanic eruptions, floods, and earthquakes. And to indicate subtle shifts in mineralogical composition. The important aspects of these studies are supported by ground truth data collected with field portable spectrometers. ASD designed the FieldSpec spectroradiometer and TerraSpec mineral analyzer systems to offer multiple configurations, resolution levels, and accessories for the collection of laboratory-quality field spectra. They feature a flexible wavelength range that covers the visible and near-infrared regions, and are ideal for measurements involving absorption and reflectance of solar radiation. Data from these highly portable field instruments is compatible with popular image analysis software packages such as ENVI, ERMapper, ERDAS Imagine and MultiSpec, allowing you to easily build focused and detailed spectral libraries tailored to your application requirements. The FieldSpec instruments are a technical tool that helps you work in sites located in some of the most remote areas of the planet.

What are the Geosciences?

The geosciences include all fields of science that deal specifically with the varying functions of the systems on earth. Geosciences can also be called earth sciences, earth science, or geoscience. The geosciences recognize earth as a unique system because it is the only planet which, to our knowledge, has life forms. While some who study the geosciences take a multi-disciplinary approach to studying our planet, others study Earth systems through a particular field of science.

The best way to understand the geosciences is to understand the underlying sciences that form this field of study. Most of these fields can be extrapolated to study systems not present on earth. However, when these fields apply specifically to studies of the earth, they are geosciences. These include studies of the atmosphere, the ocean, and plant and animal life.

One of the geosciences most important to understanding Earth is geology. By studying rock formations, differentiations in rocks, and rock development, scientists can, with increasing accuracy, date the earth, explain phenomena that affect the earth structure – such as identifying meteors that hit the earth, and make predictions about the further development of the earth and how it may impact human, animal, and plant life. More specific applications of geology may examine the operation of systems like plate tectonics or the way in which volcanoes operate, called volcanology. Additional studies may include learning more about minerals through the field of mineralogy.

The geosciences also comprise specific disciplines that study the atmosphere and how it changes. Of these, the most commonly known is meteorology, which studies and predicts weather systems. Climatology examines weather patterns over long periods of time and also evaluates how humans may impact climate, such as the human use of fossil fuels that is creating global warming. Paleoclimatology poses theories about prehistoric weather systems, largely by examining glacier materials. Atmospheric chemistry examines the chemical values that compose the earth's atmosphere, and forms another branch of study in the geosciences.

The geosciences also rely on the study of plants and animals, biology. Specializations in fields of biology include paleontology, examinations of fossil life forms which can help identify relationships between prehistoric plants, animals, and their environment. Biogeography evaluates the location of species on Earth, and studies how geographic changes in the earth impact plant and animal populations. Geomicrobiology evaluates tiny organisms or parts or organisms and their interaction with things that are not organic, such as rocks and minerals.

In the geosciences, life is not only evaluated on the ground, but is also evaluated as it exists in ocean environments. Oceanography and marine biology evaluate living systems in the ocean, and subsets of these fields may evaluate fossil records of previous ocean life or look at the way rocks and minerals form in ocean settings. Consideration is also given to the attempt to map the ocean floor, which is not completely possible at this point, given that the depths of most oceans are often inaccessible to both humans and machines.

All geosciences, combined with the principles of physics, can be used to form a holistic approach to studying the earth's systems. They can also help us to define the ways in which we may be able to increase our chances of survival by better understanding all aspects of our environment on Earth.

How do I Become a Geoscientist?

In general, geoscientists study phenomenon related to the structure, dynamics, and evolution of planet Earth. To become a geoscientist you will typically have to complete a degree at an accredited college or university. Most geoscientists have at least a Bachelor of Science degree or the equivalent. Selecting the right geoscience program could potentially increase your chances of realizing your dream to become a geoscientist. Since there are a great number of phenomena related to the study of Earth, geoscience is a vast field. It is therefore commonly subdivided into disciplines such as geology, geochemistry, and geophysics. Many geoscientists are experts in a specific subject. For example, some geoscientists may study the movement of tectonic plates, and the movement of the planet. Other such scientists investigate materials, minerals, or processes, such as rock formations, volcanoes, or lava. To become a geoscientist it can be a good idea to familiarize yourself with the field. This can be an important step in preparing yourself for your eventual studies. There are many introductory books and websites available that can offer you the basics of geoscience. Consulting these resources can illustrate for you know what are the key points of geoscience are, and also educate you on the work, and various jobs, a geoscientist may perform.

It dedicated to microwave remote sensing applications, nuclear, geophysics and telecommunication; and environment remote sensing investigations. It plays tremendous roles in building a quantitative and comprehensive understanding of how the Earth system works. The advanced remote sensing technology is also used widely to monitor and survey the natural disasters and man-made pollution.

Role of Remote Sensing

Remote sensing is the science and art of obtaining information about an object through the analysis of data acquired by a device that is not in contact with the object. Remotely sensed data can be of many forms, including variations in force distribution, acoustic wave distribution or electromagnetic energy distributions and can be obtained from a variety of platforms, including satellite, airplanes, remotely pilot vehicles, handheld radiometers or even bucket trucks. They may be gathered by different devices, including sensors, film camera, digital cameras, and video recorders. Our eyes acquire data on variations in electromagnetic radiations. Instruments capable of measuring electromagnetic radiation are called sensors. Sensors can be differentiated in two main groups:

- Passive sensors: without their own source of radiation. They are sensitive only to radiation from a natural origin.
- Active sensors: which have a built in source of radiation. Examples are Radar (Radio detection and ranging) and Lidar (Light detection and ranging) systems.

This can be analogue (photography) or digital (multispectral scanning, thermography, radar). The elements of a digital image are called resolution cells (during the data acquisition) or pixels (after the image creation). The implementation of remote sensing data by the user requires some knowledge about the technical capabilities of the various sensor systems. The technical capabilities of the sensor systems can be listed in three resolutions:

- Spatial resolution: concerns the size of the resolution cell on the ground in the direction of the flight and across. The size of the pixel determines the smallest detectable terrain feature.
- Spectral resolution: concerns the number, location in the electromagnetic spectrum and bandwidth of the specific wavelength bands or spectral bands. This resolution differs between sensors and largely determines their potential use.
- Temporal resolution: concerns the time lapse between two successive images of the same area. This primarily determined by the platform used, and secondly by the atmospheric conditions.

Remote Sensing provides spatial coverage by measurement of reflected, emitted and backscattered radiation, across a wide range of wavebands, from the earth's surface and surrounding atmosphere. Remote Sensing-based spatial coverage was measured by reflected and emitted electromagnetic radiation from the earth's surface and surrounding atmosphere. Remote sensed from data of the land surface is possible across a wide range of 9.5

Remote Sensing System Types

A. Visual Remote Sensing System The human visual system is an example of a remote sensing system in the general sense. The sensors in this example are the two types of photosensitive cells, known as the cones and the rods, at the retina of the eyes. The cones are responsible for color vision. There are three types of cones, each being sensitive to one of the red, green, and blue regions of the visi ble spectrum

Optical Remote Sensing

In Optical Remote Sensing, optical sensors detect solar radiation reflected or scattered from the earth, forming images resembling photographs taken by a camera high up in space. The wavelength region usually extends from the visible and near infrared VNIR to the short-wave infrared SWIR. Different materials such as water, soil, vegetation, buildings and roads reflect visible and infrared light in different ways. They have different colors and brightness when seen under the sun. The interpretation of optical images requires the knowledge of the spectral reflectance signatures of the various materials (natural or man-made) covering the surface of the earth. Infrared Remote Sensing Infrared remote sensing makes use of infrared sensors to detect infrared radiation emitted from the Earth's surface. The middle-wave infrared (MWIR) and long-wave infrared (LWIR) are within the thermal infrared region. These radiations are emitted from warm objects such as the Earth's surface. They are used in satellite remote sensing for measurements of the earth's land and sea surface temperature. Thermal infrared remote sensing is also often used for detection of forest fires, volcanoes, oil fires. Microwave Remote Sensing There are some remote sensing satellites which carry passive or active microwave sensors The active sensors emit pulses of microwave radiation to illuminate the areas to be imaged. Images of the earth surface are formed by measuring the microwave energy scattered by the ground or sea back to the sensors.

Satellite Remote Sensing

In this, you will see many remote sensing images acquired by earth observation satellites. These remote sensing satellites are equipped with sensors looking down to the earth. They are the "eyes in the sky" constantly observing the earth as they go round in predictable orbits. Orbital platforms collect and transmit data from different parts of the electromagnetic spectrum, which in conjunction with larger scale aerial or ground-based sensing and analysis provides researchers with enough information to monitor trends. Other uses include different areas of the earth sciences such as natural resource management, agricultural fields such as land usage and conservation, and national security and overhead, ground-based and stand-off collection on border areas.

1

CHAPTER

REMOTE SENSING STUDY OF GLACIAL CHANGE IN THE NORTHERN PATAGONIAN ICEFIELD

Lucy Dixon and Shrinidhi Ambinakudige

Keywords: Patagonia, Glaciers, South America, ASTER

INTRODUCTION

Glaciers are key indicators for assessing climate change [1-3]. Beginning in the nineteenth century, glaciers in many parts of the world began to retreated significantly, a clear indicator of warming climate [3-7]. Therefore, an accurate estimation of mass balance, glacial retreat, and glacier lake expansion in mountain glaciers will contribute to the understanding of the global impacts of climate change. Such data will also help develop an early warning system for Glacier Lake Outburst Floods (GLOF), which can devastate vulnerable downstream localities.

Further, documentation of glacier measurements in major glacial regions is not very complete. For example, European Alpine glaciers have been well documented using both in situ and remote sensing methods. The glaciers found in the Himalayans have also attracted more attention in recent decades. Conversely, South American glaciers have not been studied as intensively. Like other glacier regions, the reduction in glacier mass is also very evident in Patagonia [8]. However, documentation of the status of glaciers in the Chilean Andes has been very limited.

The glacier mass balance is the total loss or gain in a glacier's mass over the course of a hydrological year [9]. It is possible to use both indirect and direct methods to measure mass balance. The direct method involves an in-situ collection of glacier data to measure retreat and mass balance taken at lower ablation regions [10, 11]. In recent

Citation: Dixon L. and Ambinakudige S. (2015) Remote Sensing Study of Glacial Change in the Northern Patagonian Icefield. Advances in Remote Sensing, 4, 270-279. doi: 10.4236/ars.2015.44022.

years, remote sensing satellite images have helped researchers study even the world's most inaccessible glaciers [12-15]. The indirect method of mass balance measurements requires the use of remote sensing techniques, involving the use of satellite images, aerial photos, and photogrammetric methods [15, 16].

In this study, the spatial and temporal positional shifts in the terminus of the glaciers in the North Patagonian Icefield (NPI) are analysed, as well as the expansion of glacier lakes from 1979 to 2013. Following this, the mass balance of some of the major glaciers in the NPI are measured using ASTER satellite stereo images from 2007 to 2012.

STUDY AREA

The Chilean cryosphere can be divided into four parts, based on climate and glacier characteristics: 1) the Arid North (18°S to 32°S); 2) the Central Andes (32°S to 38°S), which has high precipitation in the winter and dry summers; 3) Southern Chile (38 to 41°S), which has an active volcano region; and 4) the Northern and Southern Patagonian ice fields (41°S), which have abundant precipitation [14]. The NPI and the Southern Patagonia Icefield (SPI) together have the largest temperate ice mass in the southern hemisphere [17]. The temperatures of these temperate glaciers are always at melting point throughout the glacial area and respond to climate change after short periods of time [18]. Therefore, such glaciers must be studied in order to understand the dynamics of climate change.

Patagonian glaciers have long been neglected for mass-balance studies, although researchers have undertaken other glacial studies in the area [17]. Due to the remoteness, inaccessibility, and bad weather of Patagonia, field-based mass-balance studies are lacking [18]. As glaciers in this region have not been as closely studied, researchers are uncertain about the degree of which these glacier-melts contribute to the rise of sea level [19]. Glaciers are the most prominent water source in this area; therefore, the lack of records of water balance and annual mean discharges in the region is of great concern [20]. In recent years, however, several studies have been initiated to study glacier changes in this region [21, 22], and have clearly shown glacial retreat and thinning in the Patagonian ice field.

The NPI located in the Central Andes region [23], receives abundant precipitation: approximately 2000 to 11,000 mm of water per year [19]. The eastside of NPI, however, receives precipitation of only approximately 1500 mm per year [24]. The precipitation on the eastside is predominantly attributable to mid-latitude storms that move westward between 40°S and 50°S, a00s well as the mean latitudinal locations of the storm tracks, which closely follow the axis of the westerly winds in the middle and upper troposphere [25].

The west wind becomes very strong and nearly permanent when the mean air temperature at sea level decreases gradually from 13.7°C at latitude 37°23'S to 6.5°C at latitude 53°10'S [23]. The NPI is also characterized by the Föhn winds, which are dry, warm, and down-sloping winds. These winds lead to dry conditions and high temperatures on the eastern side of the icefield. They also increase the precipitation gradient in the west-east direction [26], however in general precipitation in the NPI shows no significant seasonality [27]. The elevation of the limit between rainfall and snowfall during precipitation events is the main climatic factor that causes glacier fluctuations in the NPI [23].

The NPI is located between 46°30'S and 47°30'S, and it stretches approximately 100 km along 73°30'W, covering a total area of 4200 km^2 (Figure 1). It has a width of approximately 71 km in the west-east direction, from the tongues of the San Quintin glacier to the Soler glacier [21]. Most of the glaciers lie at altitudes of 1000 m to the west and 1500 min to the east. A northwest-to-southeast line of nunataks divides the northeast icefield from the southwest icefield. The NPI has 28 outlet glaciers [17]. The San Quintin and San Rafael glaciers are the largest glaciers in the NPI, each with an area of approximately 760 km^2, followed by the Steffan and Colonial glaciers [28]. San Rafael is the only tidewater-calving glacier in the NPI, most of the other glaciers have freshwater-calving fronts [29]. Both calving-glacier types respond directly to climate change [30].

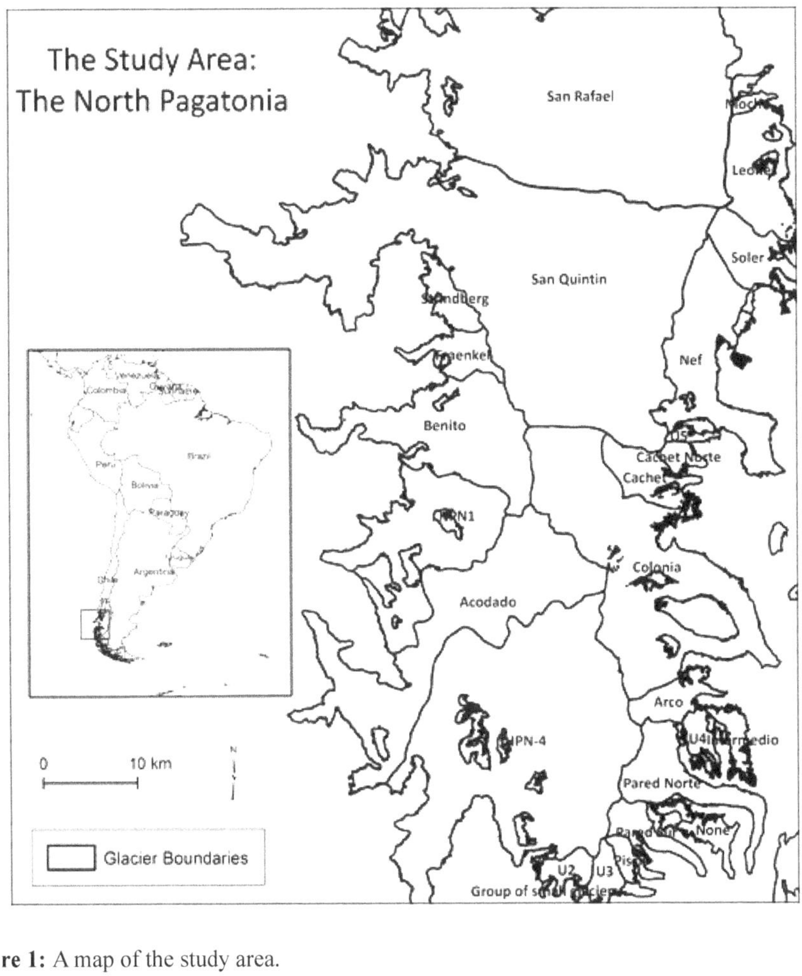

Figure 1: A map of the study area.

Equilibrium Lines of Altitude (ELA) of approximately 900 to 1350 m separate the NPI into a 2578 km^2 accumulation zone and a 1550 km^2 ablation zone [28]. The summit of Mount San Valentin, at the northeast end of the ice field, has the highest elevation, at 4032 ± 1 m above sea level. The tidal water of Laguna San Rafael, located on the western side of the ice field, is of a minimum altitude at sea level [21]. Due to climatic conditions, topography, and the nunataks that divide the icefield, the western outlet glaciers are larger and more active than the eastern glaciers [21, 31]. Over the last 60 years, [32] observed a general retreat of 21 glaciers in the NPI. After the 1990s, there was an increase in the rate of retreats, with the western side experiencing a greater retreat than the eastern side [33].

Glacier Lake Outburst Floods (GLOF) are a disastrous discharge of large volumes of water, caused by the breaking of moraine-dammed lakes [34]. Glacier lakes form from melting glacier ice and snow. Higher rates of melting have increased the threat of glacier lake outbursts in the world's high mountain ranges [35]. Various natural phenomena, such as avalanches, earthquakes, or even the weight of the water itself, break the moraine dams. Some of the most devastating GLOF occurred in the Peruvian Andes. Researchers [36] have described the natural hazards associated with the GLOF of Palcacocha Lake in Cordillera Blanca. In 2000, a rock fall partially displaced a moraine-dammed lake at the Lago Calafate glacier and caused a GLOF in the Northern Patagonia Ice Field [37].

DATA AND METHODS

Topographic maps of the study area with a scale of 1:50,000 were acquired, created by the Istituto Geografico Militar of Chile (IGM) in 1979, with a contour interval of 50 m and vertical random error of 17 m [38]. The National Imagery and Mapping Agency [39] estimated the horizontal error of the maps as 15 m. The maps were projected to WGS 1984 UTM Zone 18S.

To delineate the glaciated area from the non-glaciated area, the glacier boundaries on the ASTER 2012 and 2007 images were digitized in Arc GIS 10.1. The region that overlapped between the 2012 and 2007 images was used for the mass balance analysis. A total of 14 glaciers fell within the overlapping region: San Quintin, Fraenkel, Strindberg, Colonia, Acodado, HPN 1, HPN-4, Cachet, Cachet Norte, Nef, Arco, Benito, Pared Norte, and U5 (Figure 1).

Three Landsat images were combined with topographic maps of a scale of 1:50,000 to measure the retreat of the NPI glaciers and to analyse the expansion of glacier lakes. The Landsat images, from 1985 (Landsat 5 TM), 2000 (Landsat 7 ETM+), and 2013 (Landsat 8), were downloaded from the USGS EROS website (http://earthexplorer.usgs.gov). Both image types were re-projected to WGS 84 UTM Zone 18S. To clearly delineate the glacier lakes, the Normalized Differenced Water Index (NDWI) was calculated. In addition, the images of the lakes were visually inspected using various band combinations. The NDWI (equation 1) uses the (Near Infrared) NIR and blue band to differentiate water bodies from other features [35].

$$NDWI = \frac{B_{NIR} - B_{blue}}{B_{NIR} + B_{blue}} \quad (1)$$

Using the NDWI and the Landsat images, the glacier lakes in the NPI and in the topographic map were digitized. A significant increase in the area of the glacier lakes between 1979 and 2013 was observed.

Two ASTER images, from April 4th 2007 and March 18th 2012, were used for the mass balance estimation. ASTER satellite imagery was used due to its stereoscopic capability to generate Digital Elevation Models (DEMs) [40]. The bands 3N and 3B, with nadir and backward looking telescopes respectively, were used for the DEM creation [40]. These bands have a wavelength range of 0.78 to 0.86 μm in the NIR. As with the Landsat images, the ASTER images were georeferenced to WGS84 datum, UTM Zone 18S.

DEM Creation

The Ortho Engine extension of PCI Geomatica 2012 software was used to create the DEMs at a resolution of 15 m resolution. This involved the collection of 50 well-distributed Ground Control Points (GCPs) and 100 well-distributed Tie Points (TPs) from the 2007 and 2012 images, with priority to non-glaciated regions, as in the absence of a natural disaster, such regions hardly change. Glaciated areas, shaded areas, and cloud-covered areas were excluded from the GCP selection, as such areas are highly variable and could introduce inaccuracies to the DEM. The root mean square error (RMSE) after the GCPs collection was 2.14 m for the 2012 image, with an X RMSE of 1.81 m and a Y RMSE of 1.13 m. The RMSE for the 2007 image was 1.37 m, with an X RMSE of 1.03 m and a Y RMSE of 0.90 m. The change in elevation was computed by finding the difference in pixel-by-pixel DEM values between the 2012 the 2007 images. The elevation difference image was subsequently converted to points. Shaded areas, steep slopes, and clouded areas were excluded from the analysis. The density of ice used to calculate mass balance was fixed at 900 $kg{\cdot}m^{-3}$ [9]. The elevation of the master DEM (2007) was compared with the elevation of the 1:50,000 topographic maps. The correlation of over 70 well-distributed random points from topographic map elevations and the DEM elevations was very high (b = 0.91). Considering the lower resolution of the topographic map, DEM values were observed to be highly accurate.

The non-glaciated region was expected to have a zero-elevation difference value, as there should be no change in elevation unless there is a natural disaster or a vertical error in the DEM. The elevation differences between the 2007 and 2012 DEMs in non-glaciated areas were used to calculate the uncertainty of the mass balance estimation in the glaciated areas.

The uncertainties of the elevation difference in glacier and non-glacier areas were computed using individual standard errors (SE) [41]. The SE in the non-glaciated area was calculated using equation 2, where n is the number of included pixels:

$$SE = \frac{\text{STDV non} - \text{glac}}{\sqrt{n}} \tag{2}$$

In accordance with the law of error propagation, the SE and the mean difference of elevation (MED) of the non-glaciated area were used to calculate uncertainty (equation 3) in glaciated areas [41]:

$$e = \sqrt{(SE)^2 + (MED)^2} \tag{3}$$

RESULTS

The retreat of nine glaciers in the NPI that occurred between 1979 and 2013, San Rafael, San Quintin, Colonia, Acodado, HPN-1, Nef, Strindberg, Fraenkel, and Pared Norte were measured, as shown in Figure 2. The retreat was measured via the change in position of the tongue of the glacier during the study period [21] (Figure 2).

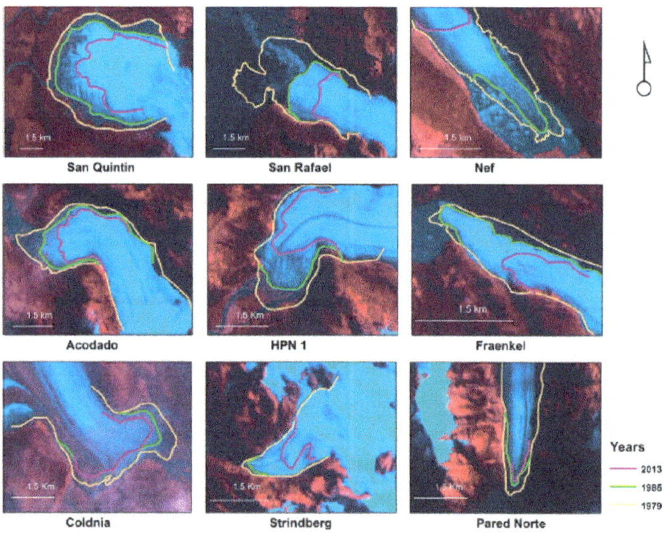

Figure 2. Retreat of nine major glaciers in the NPI.

Table 1 shows the extent of the retreat of all nine glaciers, with San Quintin retreating the most, at 4.5 km. The tongue of this glacier was intact in 1979, however by 2013 had retreated to form a new lake. Another glacier tongue in San Quintin retreated by 3.03 km. San Rafael retreated by approximately 3.3 km between 1979 and 1987 [21]. San Quintin, conversely, retreated by only approximately 0.5 km within the same time period. In general, all glaciers in the study area retreated, which can be attributed to atmospheric warming within the region [21]. The rate at which these glaciers retreat is a direct response to climate as these are temperate glaciers that respond rapidly to changes in temperature [23]. Due to the lack of weather data for the region, however, it was not possible to analyse long-term temperature changes.

Table 1. Retreat of glaciers between 1979 and 2013

Glacier	Retreat (km)
San Quintin	4.50
San Rafael	3.92
Nef	3.82
Acodado	3.40
HPN-1	2.43
Fraenkel	2.19
Colonia	2.13
Strindberg	1.48
Pared Norte	0.83

The number of glacier lakes in the study area increased significantly between 1979 and 2013. In 1979 there were nine glacier lakes; in 1985 there were 10; in 2000 this rose to 11; and by 2013 the glacier lakes reached 13. The glacier lake area increased from 13.49 km^2 in 1979 to 65.06 km^2 in 2013 (**Figure 3**).

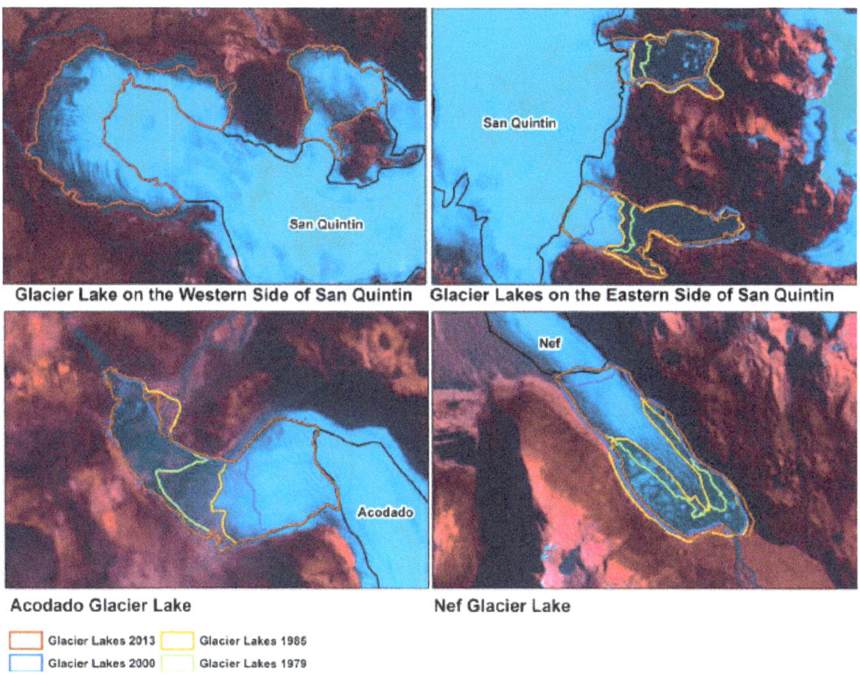

Figure 3. Expansion of glacier lakes.

The maximum expansion of area was observed for San Quintin glacier lake; from 5.42 km^2 to 42.37 km^2. This expansion is the result of the creation of a new glacier lake at the tongue of San Quintin. A study [42] observed a similar trend between 1945 and 2011, with such a drastic increase comparable to other glaciers. **Table 2** shows the number of lakes and their total area.

Table 2. Glacier lakes area for 1979 and 2013

Glacier Lake	1979 Area (km^2)	2013 Area (km^2)
San Quintin	5.42	42.37
Fraenkel	1.48	2.41
Benito	0.39	1.00
Acodado	1.80	9.38
Cachet	2.50	4.30
Nef	1.90	5.60

In the non-glaciated area, 522,513 random points were selected to estimate the uncertainty in the mass balance estimation. The average elevation difference in the non-glacial area was 2.82 m, with a standard deviation of 8.99 m and a standard error of mean of 0.012 m. Based on these parameters, uncertainty in the mass balance of the glaciated area was estimated at ±0.51 m.w.e.a^{-1}. The mass balance for accumulation and ablation zones for the glaciers were estimated separately. **Table 3** presents mass balance values from 2007 to 2012.

To estimate the mass balance, the areas of glaciers that fell within the boundaries of the satellite images were used. In general, ablation zones had higher loss of mass compared accumulation zones (**Table 3**). The HPN1 glacier ablation zone showed the highest loss of mass, with −4.76 ± 0.51 m.w.e.a^{-1}, followed by Pared Norte (−4.58 ± 0.51 m.w.e.a^{-1}), the Strindberg ablation zone (−3.28 ± 0.51 m.w.e.a^{-1}), the Acodado ablation zone (−3.00 ± 0.51 m.w.e.a^{-1}), the Nef ablation zone (−2.33 ± 0.51 m.w.e.a^{-1}), the Fraenkel ablation zone (−1.86 ± 0.51 m.w.e.a^{-1}), the San Quintin ablation zone (−1.80 ± 0.51 m.w.e.a^{-1}), the Nef accumulation zone (−1.77 ± 0.51 m.w.e.a^{-1}), the Colonia ablation zone (−1.52 ± 0.51 m.w.e.a^{-1}), the HPN-4 ablation zone (−1.27 ± 0.51m.w.e.a^{-1}), the HPN-4 accumulation zone (−1.25 ± 0.51 m.w.e.a^{-1}), Acodado (−1.21 ± 0.51 m.w.e.a^{-1}), and the Benito ablation zone (−1.14 ± 0.51 m.w.e.a^{-1}). Accumulation zones of HPN1 (−0.89 ± 0.51 m.w.e.a^{-1}), Strindberg (−0.74 ± 0.51 m.w.e.a^{-1}), San Quintin (−0.72 ± 0.51 m.w.e.a^{-1}), and Fraenkel (−0.23 ± 0.51 m.w.e.a^{-1}) also showed negative mass balance, however with values close to uncertainty values. Accumulation zones in Cachet Norte, Arco, and U5 glaciers had positive mass changes; other accumulation zones also showed positive mass changes, but the estimates were close to the uncertainty in the model estimation.

Table 3. Mass balance of glaciers between 2007 and 2012. Note that "Area" is not the actual area of the glaciers, but represents only the area that was covered in the image

Glaciers	*Area* (km^2)	*Average Elevation Difference* ($m.a^{-1}$)	*STD* ($m.a^{-1}$)	*Specific Mass Balance* ($m.w.e.a^{-1}$)
Acodado Accumulation	190.96	−1.21	3.14	−1.09 ± 0.51
Acodado Ablation	44.82	−3.00	4.15	−2.70 ± 0.51
Arco	10.31	0.84	3.19	0.75 ± 0.51
Benito-Accumulation	108.91	0.43	2.51	0.38 ± 0.51
Benito Ablation	71.12	−1.14	2.26	−1.03 ± 0.51
Cachet	46.86	0.54	3.90	0.49 ± 0.51
Cachet Norte accumulation	10.75	1.38	3.51	1.24 ± 0.51
Cachet Norte Ablation	1.04	0.14	2.50	0.13 ± 0.51
Colonia Accumulation	246.50	0.08	3.61	0.07 ± 0.51
Colonia Ablation	24.47	−1.52	3.48	−1.37 ± 0.51
Fraenkel Accumulation	24.99	−0.23	2.35	−0.21 ± 0.51
Fraenkel Ablation	9.74	−1.86	2.79	−1.67 ± 0.51
HPN1 Accumulation	102.31	−0.89	2.79	−0.80 ± 0.51
HPN1 Ablation	67.62	−4.76	3.48	−4.28 ± 0.51
HPN-4 Accumulation	264.55	−1.25	2.94	−1.13 ± 0.51
HPN-4 Ablation	5.98	−1.27	3.41	−1.15 ± 0.51
Nef Accumulation	86.52	−1.77	4.57	−1.59 ± 0.51
Nef Ablation	7.54	−2.33	3.04	−2.09 ± 0.51
Pared Norte Accumulation	5.19	−4.58	7.47	−4.13 ± 0.51
San Quintin Accumulation	410.10	−0.72	2.99	−0.65 ± 0.51
San Quintin Ablation	107.81	−1.80	3.00	−1.62 ± 0.51
Strindberg Accumulation	23.83	−0.74	1.90	−0.66 ± 0.51
Strindberg Ablation	3.24	−3.28	2.38	−2.95 ± 0.51
U5	6.63	0.68	3.42	0.61 ± 0.51

DISCUSSION AND CONCLUSIONS

The results of this study clearly indicated the general retreat and expansion of glacier lakes, and the loss of mass in the NPI. However, other studies have found the advancement of glacial tongues in various glaciers in the previous decade. For example, the San Quintin glacier advanced in 1993 and subsequently retreated between 1996 and 2000 [43]. Our study showed that the retreat in San Quintin is continuing. Overall, there was a 4.5 km retreat in the San Quintin glacier tongue between 1979 and 2013. A prior study observed that the retreat of the San Rafael glacier stopped in 1991 [44]. However, the same study found a 3.92 km retreat in the San Rafael glacier during the study period.

A study [29] observed a mass balance loss in all the glaciers in the NPI between the years of 1975 and 2001. These average ice-loss rates have more than doubled to an equivalent sea level rise in the North and South Patagonian Ice fields between 1995 and 2000 [19]. The same study observed a negative mass balance in most glaciers. This study also found that accumulation zones are losing mass at a higher rate than ablation zones.

Glaciers on the western side of the nunataks experienced more retreat than glaciers on the eastern side of the nunataks. Another study [28] referred to orographic factors as the cause of this pattern, with the eastern side of the NPI receiving less snowfall compared to the western side, as the NPI lies in the zone of the Westerlies. The eastern side thus receives more snow from the western side through drift snow blown by wind. Due to this drift snow, the eastern side accumulates more snow and experiences less retreat in the NPI. A study [45] estimated the thinning rate of the HPN-1 glacier as -0.7 m.w.e.a^{-1} between 1945 and 1975; another study [29] estimated the thinning to be -4.0 m.w.e.a^{-1} between 1975 and 2001; and we recorded a negative mass loss of HPN-1 at -4.76 ± 0.51 m.w.e.a^{-1} in the ablation zone and -0.80 ± 0.51 m.w.e.a^{-1} in the accumulation zone.

Although these results were not validated with *in-situ* data for either mass balance or retreat, similar results were observed by [18] [19] [28] [45] for mass balance studies and [21] [29] [41] for glacier retreating and lakes. The results of this study, as well as previous studies, show an overwhelming retreat of glaciers and loss of mass in the NPI. These results must be taken seriously, as these conditions are linked to climate fluctuations within the region.

REFERENCES

1. Ambinakudige, S. (2014) Glaciers. In: Warf, B., Ed., *Oxford Bibliographies in Geography*, Oxford University Press, New York.
2. Bolch, T., Buchroithner, M., Pieczonka, T. and Kunert, T. (2008) Planimetric and Volumetric Glacier Changes in Khumbu Himalaya since 1962 Using Corona Landsat TM and ASTER Data. *Journal of Glaciology*, **54**, 592-600. http://dx.doi.org/10.3189/002214308786570782
3. IPCC. (2013) Summary for Policymakers. In: Stocker, T.F., Qin, D., Plattner, G.-K., Tignor, M., Allen, S.K., Boschung, J., Nauels, A., Xia, Y., Bex, V. and Midgley, P.M., Eds., *Climate Change* 2013: *The Physical Science Basis. Contribution of Working Group I to the Fifth Assessment Report of the Intergovernmental Panel on Climate Change*, Cambridge University Press, Cambridge, UK and New York, USA.
4. Ambinakudige, S. (2010) A Study of the Gangotri Glacier Retreat in the Himalayas Using Landsat Satellite Images. *International Journal of Geoinformatics*, **6**, 7-12.
5. Ambinakudige, S. and Joshi, K. (2015) Multi-Decadal Changes in Glacial Parameters of the Fedchenko Glacier in Tajikistan. *International Journal of Advanced Remote Sensing and GIS*, **4**, 911-919.
6. Barry, R.G. (2006) The Status of Research on Glaciers and Global Glacier Recession: A Review. *Progress in Physical Geography*, **30**, 285-306. http://dx.doi.org/10.1191/0309133306pp478ra
7. Liu, K.C., Reese, A. and Thompson, L.G. (2005) Ice-Core Pollen Record of Climatic Changes in the Central Andes during the Last 400 yr. *Quaternary Research*, **64**, 272-278. http://dx.doi.org/10.1016/j.yqres.2005.06.001

8. Casassa, G., Haeberli, W., Jones, G., *et al.* (2007) Current Status of Andean Glaciers. *Global and Planetary Change*, **59**, 1-9. http://dx.doi.org/10.1016/j.gloplacha.2006.11.013

9. Paterson, W. (1994) The Physics of Glaciers. 3rd Edition, Pergamon, Oxford.

10. Nishida, K., Satow, K. Aniya, M., Casassa, G. and Kadota, T. (1995) Thickness Change and Flow of Tyndall Glacier Patagonia. *Bulletin of Glacier Research*, **13**, 29-34.

11. Rott, H., Rack, W., Skvarca, P. and De Angelis, H. (2002) Northern Larsen Ice Shelf, Antarctica: Further Retreat after Collapse. *Annals of Glaciology*, **34**, 277-282. http://dx.doi.org/10.3189/172756402781817716

12. Frey, H., Machguth, H., Huss, M., Huggel, C., Bajracharya, S., Bolch, T., Kulkarni, A., Linsbauer, A., Salzmann, N. and Stoffel, M. (2014) Estimating the Volume of Glaciers in the Himalayan-Karakoram Region Using Different Methods. *The Cryosphere*, **8**, 2313-2333. http://dx.doi.org/10.5194/tc-8-2313-2014

13. Bhambri, R., Bolch, T., Chaujar, R. and Kulshreshtha, S. (2011) Glacier Changes in the Garhwal Himalaya, India, from 1968 to 2006 Based on Remote Sensing. *Journal of Glaciology*, **57**, 543-556. http://dx.doi.org/10.3189/002214311796905604

14. Pellicciotti, F., Ragettli, S., Carenzo, M. and McPhee, J. (2014) Changes of Glaciers in the Andes of Chile and Priorities for Future Work. *Science of the Total Environment*, **493**, 1197-1210. http://dx.doi.org/10.1016/j.scitotenv.2013.10.055

15. Racoviteanu, A., Arnaud, Y., Williams, M. and Ordonez, J. (2008) Decadal Changes in Glacier Parameters in the Cordillera Blanca, Peru Derived from Remote Sensing. *Journal of Glaciology*, **54**, 499-510. http://dx.doi.org/10.3189/002214308785836922

16. Oerlemans, J. (2005) Extracting a Climate Signal from Glacier Records. *Science*, **308**, 675-677. http://dx.doi.org/10.1126/science.1107046

17. Warren, C.R. and Sugden, D.E. (1993) The Patagonian Icefields: A Glaciological Review. *Arctic and Alpine Research*, **25**, 316-331. http://dx.doi.org/10.2307/1551915

18. Aniya, M., Sato, H., Naruse, R., Skvarca, P. and Casassa, G. (1996) The Use of Satellite and Airborne Imagery to Inventory Outlet Glaciers of the Southern Patagonia Icefield South America. *Photogrammetric Engineering and Remote Sensing*, **62**, 1361-1369.

19. Rignot, E., Rivera, A. and Casassa, G. (2003) Contribution of the Patagonia Icefields of South America to Sea Level Rise. *Science*, **302**, 434-437. http://dx.doi.org/10.1126/science.1087393

20. Escobar, F., Vidal, F. and Garin, C. (1992) Water Balance in the Patagonia Icefield. In: Natuse, R., Ed., *Glaciological Researches in Patagonia*, Japanese Society of Snow and Ice, Nagoya, 109-119.

21. Lopez, P., Chevallier, P., Favier, V., Pouyaud, B., Ordenes, F. and Oerlemans, J. (2010) A Regional View of Fluctuations in Glacier Length in Southern South

America. *Global and Planetary Change*, **71**, 85-108. http://dx.doi.org/10.1016/j.gloplacha.2009.12.009

22. Willis, M.J., Melkonian, A.K., Pritchard, M.E. and Ramage, J.M. (2012) Ice Loss Rates at the Northern Patagonian Icefield Derived Using a Decade of Satellite. *Remote Sensing of Environment*, **117**, 184-198. http://dx.doi.org/10.1016/j.rse.2011.09.017
23. Lliboutry, L. (1998) Glaciers of Wet Andes. In: Ferrigno, W.M.J., Ed., *Satellite Image Atlas of Glaciers of the World*, US Government Printing Office, Washington DC, I109-I206.
24. Enomoto, H. and Nakajima, C. (1985) Recent Climate-Fluctuations in Patagonia. In: Nakajima, C., Ed., *Glaciological Studies in Patagonia Northern Icefield* 1983-1984, Data Center for Glacier Research, Japanese Society of Snow and Ice, Nagoya, 7-14.
25. Garreaud, R., Vuille, M., Campagnucci, R. and Marengo, J. (2009) Present-Day South American Climate. *Paleogeography, Paleoclimatology, Paleoecology*, **281**, 180-195. http://dx.doi.org/10.1016/j.palaeo.2007.10.032
26. Fujiyoshi, Y., Kondo, H., Inoue, J. and Yamada, T. (1987) Characteristics of Precipitation and Vertical Structure of Air Temperature in the Northern Patagonia. *Bulletin of Glacier Research*, **4**, 15-23.
27. Carrasco, J., Casassa, G. and Rivera, A. (2002) Meteorological and Climatological Aspects of the Southern Patagonia Icefield. In: Casassa, G., Sepulveda, F. and Sinclair, R., Eds., *The Patagonian Icefields: A Unique Natural Laboratory for Environmental and Climate Change Studies*, Kluwer Academic/Plenum Publishers, New York, 29-65. http://dx.doi.org/10.1007/978-1-4615-0645-4_4
28. Aniya, M. (1988) Glacier Inventory for the Northern Patagonia Icefield Chile and Variations 1944/45 to 1985/86. *Arctic and Alpine Research*, **20**, 179-187. http://dx.doi.org/10.2307/1551496
29. Rivera, A., Benham, T., Casassa, G., Bamber, J. and Dowdeswell, J. (2007) Ice Elevation and Areal Changes of Glaciers from the Northern Patagonia Ice Field Chile. *Global and Planetary Change*, **59**, 126-137. http://dx.doi.org/10.1016/j.gloplacha.2006.11.037
30. Warren, C. and Aniya, M. (1999) The Calving Glaciers of Southern South America. *Global and Planetary Change*, **22**, 59-77. http://dx.doi.org/10.1016/S0921-8181(99)00026-0
31. Aniya, M. and Enomoto, H. (1986) Glacier Variations and Their Causes in the Northern Patagonia Icefield Chile since 1944. *Arctic and Alpine Research*, **18**, 307-316. http://dx.doi.org/10.2307/1550888
32. Aniya, M. (2007) Glacier Variations of Hielo Patagonico Norte, Chile for 1944/45-2004/05. *Bulletin of Glaciological Research*, **24**, 59-70.
33. Aniya, M. (1992) Glacier Variation in the Northern Patagonia Icefield Chile between 1985/86 and 1990/91. *Bulletin of Glaciological Research*, **10**, 83-90.

34. [34] Yamanda, T. and Sharma, C. (1993) Glacier Lakes Outburst Floods in the Nepal Himalaya. In: Young, G.J., Ed., *Snow and Glacier Hydrology* (*Proceedings of an International Symposium*, *Kathmandu*, 16-21 *November* 1992), IAHS No. 218, 319-330.

35. Huggel, C., Kaab, A., Haeberli, W., Teysseire, P. and Paul, F. (2002) Remote Sensing Based Assessment of Hazards from Glacier Lake Outbursts: A Case Study in the Swiss Alps. *Canadian Geotechnical Journal*, **39**, 316-330. http://dx.doi.org/10.1139/t01-099

36. Vilımek, V., Zapata, M.L., Klimes, J., Patzelt, Z. and Santillan, N. (2005) Influence of Glacial Retreat on Natural Hazards of the Palcacocha Lake Area, Peru. *Landslides*, **2**, 107-115. http://dx.doi.org/10.1007/s10346-005-0052-6

37. Harrison, S., Glasser, N., Winchester, V., Haresign, E., Warren, C. and Jansson, K. (2006) A Glacial Lake Outburst Flood Associated with Recent Mountain Glacier retreat, Patagonian Andes. *The Holocene*, **16**, 611-620. http://dx.doi.org/10.1191/0959683606hl957rr

38. Falkner, E. (1995) Aerial Mapping: Methods and Applications. CRC Press Inc., Boca Raton, 322.

39. NIMA (1997) Department of Defense World Geodetic System 1984: Its Definition and Relationships with Local Geodetic Systems. Third Edition, NIMA TR8350.2, National Imagery and Mapping Agency, Bethesda, MD.

40. Hirano, A., Roy, W. and Lang, H. (2003) Mapping from ASTER Stereo Image Data: DEM Validation and Accuracy Assessment. *ISPRS Journal of Photogrammetry and Remote Sensing*, **57**, 356-370. http://dx.doi.org/10.1016/S0924-2716(02)00164-8

41. Bolch, T., Pieczonka, T. and Benn, D.I. (2011) Multi-Decadal Mass Loss of Glaciers in the Everest Area (Nepal Himalaya) Derived from Stereo Imagery. *The Cryosphere*, **5**, 349-358. http://dx.doi.org/10.5194/tc-5-349-2011

42. Loriaux, T. and Casassa, G. (2012) Evolution of Glacial Lakes from the Northern Patagonia Icefield and Terrestrial Water Storage in a Sea-Level Rise Context. *Global and Planetary Change*, **102**, 33-40. http://dx.doi.org/10.1016/j.gloplacha.2012.12.012

43. Harrison, S., Warren, R.C.R., Winchester, V. and Aniya, M. (2001) Onset of Rapid Calving and Retreat of Glacier San Quintin, Hielo Patagónico Norte, Southern Chile. *Polar Geography*, **25**, 54-61. http://dx.doi.org/10.1080/10889370109377705

44. Winchester, V. and Harrison, S. (1996) Recent Oscillations of the San Quintín and San Rafael Glaciers Patagonian Chile. *Geografiska Annaler*, **78**, 35-49.

45. Aniya, M. (1999) Recent Glacier Variations of the Hielos Patagonicos South America and Their Contribution to Sea- Level Change. *Arctic*, *Antarctic and Alpine Research*, **31**, 165-173. http://dx.doi.org/10.2307/1552604

2
CHAPTER

SNOW DEPTH RETRIEVAL WITH UAS USING PHOTOGRAMMETRIC TECHNIQUES

Benjamin Vander Jagt, Arko Lucieer, Luke Wallace, Darren Turner and Michael Durand

Keywords: Photogrammetry, cryosphere, unmanned aerial systems

INTRODUCTION

The Need for Accurate Depth Estimates of Snow Depths and Existing Methods

Major areas in the world depend on snow and snowmelt for the majority of their water needs [1]. Snowmelt runoff is used for crop irrigation, industrial and manufacturing processes, municipal water supply, and recreational purposes. Additionally, snowmelt has been found to impact carbon uptake from vegetated areas, thus playing a role in climate variability [2]. Being able to provide accurate and reliable estimates of snow and its water equivalent (SWE) is required for hydrologic modelling, flood prediction, and risk mitigation.

Efforts to measure snow depth and water equivalent have an extensive historical record, starting with in situ methods (e.g., [3,4]). In situ sampling methods such as snow courses, snow pillows, and snow pits, while accurate, often only provide point estimates of a spatially varying quantity. Spatial interpolation of in situ measurements has been investigated; however even the most accurate methods (e.g., kriging) only explain 30 percent of the overall variability in snow depth [5]. Additionally, in situ methods are subject to logistical and safety restrictions that limit when and where measurements can be made, especially in mountainous areas which are often characterized by high

Citation: Vander Jagt, B.; Lucieer, A.; Wallace, L.; Turner, D.; Durand, M. Snow Depth Retrieval with UAS Using Photogrammetric Techniques. Geosciences 2015, 5, 264-285. doi:10.3390/geosciences5030264.

snowfall. Because of these limitations, there exists a need for direct observations of snow depth with improved spatial coverage and reduced personnel risk compared to existing conventional methods.

Remote sensing provides a means by which measurements can be made in a temporal and spatially distributed manner, while at the same time minimizing the risk associated with traditional in situ based methods. Historically, most space-borne and airborne remote sensing methods for SWE retrieval have relied on passive microwave (PM) observations, which are sensitive to the overall mass of snow contained within the microwave footprint [6–8]. The recent GlobSnow product [9] has produced global maps of SWE. However, due to the challenges posed by vegetation, spatial heterogeneity, and deep snow, GlobSnow does not estimate SWE in mountainous areas. Forested and alpine areas where vegetation and deep snow pack interferes with the PM signal have posed significant challenges for PM-based approaches [10]. Thus, measurement methods that are capable of accurately characterizing and quantifying the snow properties in such environments are desirable as a way to augment the existing coarse resolution PM observations, which typically have pixel resolutions on the order of 10–25 km^2 at frequencies used for snow retrieval.

Deems et al. (2013) survey recent progress in development of LiDAR-based snow depth measurements from manned airborne remote sensing techniques. These have the potential to provide both high-accuracy and high-resolution estimates of snow depth with sub-meter sampling intervals and decimetre level accuracy [11]. However, monitoring snow depth with manned airborne platforms is potentially prohibitively expensive, due to the costs associated with equipping, flying, and maintaining an aircraft outfitted with a mapping quality LiDAR sensor. Thus, a method that produces data with the resolution and accuracy of conventional LiDAR, but with reduced cost would be well received within the snow science community, and is therefore the main motivation of this study.

The objectives of this study are threefold. First, to describe a novel technique for remotely measuring high-resolution spatially continuous snow depths using UAS-based photogrammetry. Second, to accurately estimate snow depths using a low-cost platform with commercial off the shelf (COTS) components. Lastly, to determine the extent to which snow depth information can accurately be obtained using airborne GPS aerotriangulation, compared to conventional triangulation techniques that incorporate ground control points (GCPs) in the adjustment.

The paper is organized into four sections. The remainder of the first section describes the increasing use of UAS in scientific applications, as well as previous efforts that have been made to measure snow depth using photogrammetric methods. In the second section, we describe the study area, UAS, and COTS sensor components that were used in this study. The third section explains in detail the two bundle adjustment workflows employed herein, as well the method by which dense point clouds are generated from overlapping imagery. In the fourth section, we present the results, including snow depth estimates generated from both methods, and suggest ways in which our methodology can be improved.

Use of Micro Unmanned Aerial Systems (UAS)

The use of UAS for conducting scientific research is continuing to grow at a rapid pace [12]. There are many reasons for this, including the decreased cost of operating and maintaining a UAS for data collection compared to a space-borne or manned airborne platform, as well as the ease of deployment for time-sensitive campaigns. While current UAS technology is not capable of the same spatial and/or temporal coverage as space-borne or manned airborne platforms, UASs provide high-resolution datasets and create new opportunities for scientific measurement at local scales.

Current research into the use of UAS as a 3D data-capture platform includes archaeological surveys, landslide deformation, glacier dynamics, and vegetation monitoring [13–16]. With the miniaturization of navigation sensors and laser scanners in recent years, it is now possible to integrate a complete airborne LiDAR mapping system that weighs less than three kilograms [17], and produces data accuracy suitable for scientific studies.

Photogrammetric methodologies have enjoyed renewed scientific interest following the recent proliferation of UAS [12,18]. These studies make use of image matching techniques in conjunction with optimized bundle adjustment (BA) routines, and allow for high-density 3D point clouds to be generated from the multi-view imagery collected by UAS. Due to the lower altitude at which UAS are typically operated, 3D scene reconstruction can be performed with increased resolution as compared to conventional airborne platforms. This increased resolution has facilitated certain forestry metrics to be measured with increased accuracy and ease [19], and thus lends credence to our study methodology.

To our knowledge, this is the first attempt to use imagery obtained from UAS in conjunction with photogrammetric techniques to resolve spatially continuous snow depths.

Prior Photogrammetric Efforts for Snow Depth Retrieval

Previous efforts have been made to map snow surface using photogrammetric techniques from imagery captured by airborne platforms; the earliest and most notable of which were [20,21]. Photogrammetric techniques cannot measure snow depth directly, rather they are used to construct two-point cloud datasets; one of which corresponds to the ground (typically observed in fall), and the other/s during the winter snow accumulation season. The points are subsequently differenced to yield snow depths. Both [20,21] relied on the physical installation of ground control points (in the form of wooden posts) throughout the study area to properly orient the stereo model generated from multiple image pairs. Additionally, the snow depth was only determined at discrete locations manually delineated by the photogrammetric analyst that processed the imagery. Subsequent studies [22] have also used networks of ground control points for the computation of the snow surface. However, the accuracy of snow depth retrieval in those studies has not reached the level achieved in [20] or [21]; both of which were characterized by considerable error. The snow depths measured in [20] exhibited a consistent bias of ~23 cm, with a 1σ of 23.7 cm. Similarly, after the removal of a mean bias of ~15 cm from the differenced surface points, [21] was able to resolve snow depth with a 1σ precision of 21 cm. It should be noted that both the [20,21] studies were completed in the years of 1965 and 1967, respectively, and to their credit, are still considered the benchmark from which to improve upon. At the time of this writing, another manuscript is also currently in review [23] that utilizes imagery taken with COTS cameras onboard a manned aircraft to map spatially continuous snow depth over an Alaskan field transect. The results are not published at the time of writing, but are the modern follow-up to [20,21] and look to be very encouraging.

As described above, all previous photogrammetric studies have relied on the use of photo-identifiable ground control points (GCPs) distributed throughout the study area to determine the position and orientation parameters of the camera centre through Bundle Adjustment (BA) prior to multi-ray intersection for surface computation. Because this represents significant labour efforts in the field, it arguably negates the benefits of using a remote sensing methodology. In this study, we examine the feasibility of airborne GPS aerotriangulation, in addition to conventional GCP-based techniques. With airborne aero-triangulation, only the trajectory of the platform's navigation solution is used to provide positional and stochastic constraints to the adjustment. Intuitively, this means that control points are measured in the sky during the UAS flight, rather than on the ground. There has been a considerable amount of published literature on this topic, and an excellent overview can be found in [24]. However, most previous efforts have been made with metric cameras and expensive, high-accuracy inertial navigation systems [25,26], whereas we rely on consumer grade COTS digital cameras for this study. The benefits of a direct photogrammetric approach are clear; if snow depth can be accurately resolved without GCP information, areas could be directly mapped without the need to expose field personnel to dangerous weather conditions.

STUDY AREA AND DATA

The chosen study site was an alpine transect within Mount Field National Park, located in south central Tasmania, Australia (see Figure 1). The study site was located at 42.68°S, 146.6°E, at an elevation of approximately 1150 m above sea level. Because mountainous environments are characterized by variable and sloped terrain, they present unique challenges for photogrammetric measurements, in contrast to planar surfaces [27]. A strong gradient in elevation, as well as thick vegetation and various soil/rock types characterized our study area.

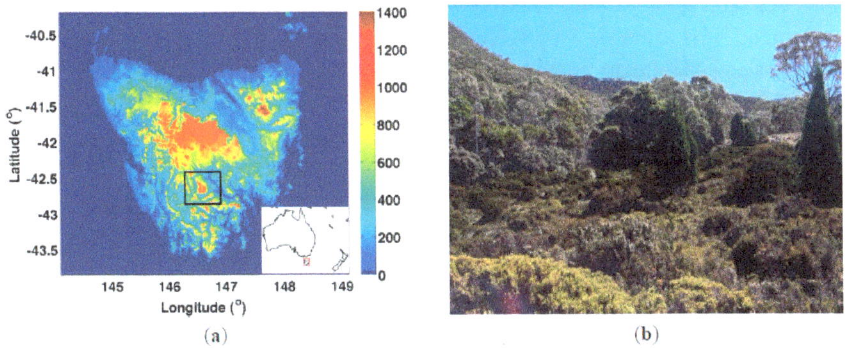

Figure 1. (**a**) Elevation map of the Mt Field study area in south central Tasmania, with inset of Australia; (**b**) Dense vegetation and variable terrain were characteristic of our study area.

Because snowfall and subsequent accumulation is quite unpredictable in Tasmania, the snow-covered dataset was collected after the first major snowfall of the season, which occurred in mid-July 2013. The dataset was collected on the eastern (leeward) side of Mt. Mawson, due to favourable weather conditions for UAS flights. The snow-free dataset was collected on 5 February 2014, after snow ablation. The total size of the study area was approximately 6900 m^2, or 0.007 km^2. A total of 31 ground control targets were spread throughout the scene, in order to provide independent validation of the photogrammetric measurements. Following the methods of [28], we used square targets (45 × 45 cm) for the validation. The ground control was surveyed using dual-frequency differential GPS observations, with coordinate accuracies in the range of 2–4 cm [29].

We measured snow depth by observing GPS positions at the snow surface, then extending the survey pole to the ground surface, and observing the position once more. A total of 37 observations of snow depth were made throughout the study area for validation of our methodology. The snow depths ranged from 33 to 121 cm, with a mean and standard deviation of 61.8 cm and 24.4 cm, respectively. A small number of measurements were in several wind drifts, which had depths in excess of 1 m (Figure 2). As will be described in Section 3.2, the technique by which the in situ snow depth measurements were made is problematic due to unseen vegetation below the snow surface.

The UAS that was used in the study is an electric powered multi-rotor platform (Droidworx SkyJiB). The octocopter is being developed by the TerraLuma research group at the University of Tasmania and was modified to accommodate the sensor suite that was used in this study (Figure 3). Multi-rotor UAS offer an increased stability and manoeuvrability in comparison to other platforms such as fixed wings or single rotor UAS, as well as low vibration. To reduce blur in the images, the camera payload is mounted on a rigid frame that is isolated from the motor vibration through the use of four silicon mounts. The UAS has eight brushless motors which operate at different rotor speeds to achieve directional flight. The total sensor payload was ~2.4 kg. A total of three flights were flown, to ensure adequate data collection, with varying flight lengths between 2.6–3.1 min. The flying height above the terrain varied considerably, but averaged ~29 m, with a corresponding GSD of 6 mm. To minimize the effects of shadows in subsequent processing, the flights were collected at between 11.00 and 14.00 local time.

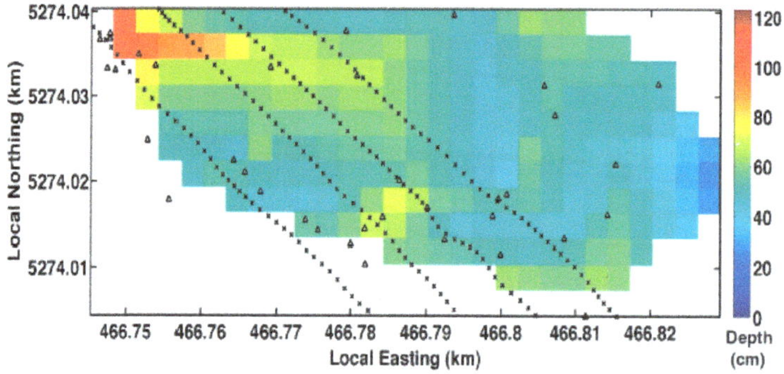

Figure 2. GPS measured snow depths from the study area, interpolated to a uniform grid at 3 m resolution. The black dots indicate image exposure locations along the unmanned aircraft system (UAS) flight trajectory, whereas the black triangles indicate measured snow depths.

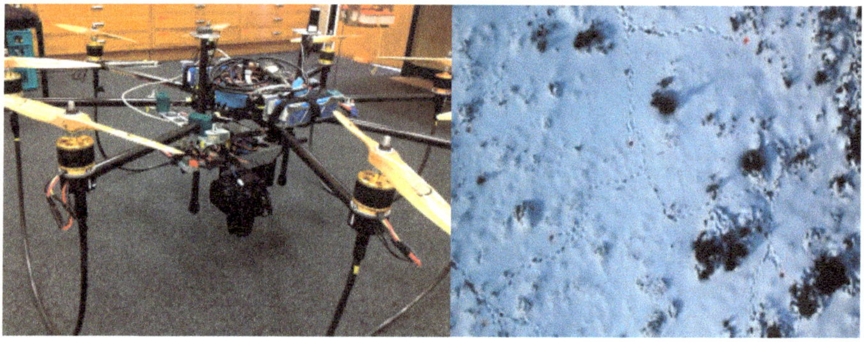

Figure 3. The UAS used in this study is shown (**left**); along with an example image (**right**) that was characteristic of our dataset. Footprints in the snow provide a sense of scale.

For image collection, we used a Canon 550D DSLR camera with a fixed 20-mm lens. The camera recorded 18-megapixel full image frames at an approximate rate of 3.8 Hz. Given the altitude (~30–45 m) and flight speed of the UAS, such a dense sampling of images is not needed. To improve computation times, we reduced the amount of photos used in the bundle adjustment by 50%. While there were slight variations in overlap, forward overlap generally ranged from 85% to 90%, while sidelap ranged from 50% to 75%. Figure 3 shows an example image from our study area during the snow-covered flight. The image frames were synched to GPS time using the camera flash to send an event trigger to the GPS receiver, to which the time information was ascribed [28]. The camera was calibrated prior to flight using standard methods. While the interior orientation parameters of the camera are often included as additional unknowns in the bundle adjustment, the camera and lens was calibrated with sufficient accuracy to carry the interior orientation parameters as observations with small measurement uncertainty.

GPS observations of camera position were recorded at 20 Hz at both L1 and L2 frequencies, using a Novatel OEMV1-df receiver. A fixed ambiguity phase solution was obtained using differential processing, with predicted coordinate accuracies of 2–3 cm. These are predicted values, as it is impossible to validate the accuracy of the solution while the platform is in motion. Inertial measurement unit (IMU) data was collected using the Microstrain 3DM-GX3 35 IMU, which records angular rate and acceleration at a rate of 100 Hz. The acceleration and angular rate data streams were time-synchronized with an internal GPS receiver, thus providing the position and orientation data needed for the bundle adjustment methods discussed below. The fusion of the IMU data with observations from the GPS receiver was carried out within a Sigma Point Kalman Smoother, in order to overcome the presence of large orientation errors which are prevalent when using MEMS-IMU sensors [17]. The state-based estimator used in this work is the Square Root Unscented variant of the Sigma Point Kalman filter (SPKF) outlined in [30]. For more specific details on the fusion algorithm, we refer readers to [17].

One of the benefits of UAS is their affordability and accessibility via COTS sensors. A listing of the platform/sensors used in this study and their corresponding price is presented in Table 1. We are not advocating the use of the manufactures that are listed in Table 1, rather it is only meant to serve as a general cost outline for such a system.

Table 1. Price comparison listing of the components used in this study

Platform/Sensor	**Indirect Bundle Adjustment**	**Price (USD)**
	Manufacturer/Model	
UAS	Droidworx SkyJiB	
Camera	Canon 550D DSLR	
	Total Cost	**$10,900**
UAS	Droidworx SkyJiB	
Camera	Canon 550D DSLR	
IMU	Microstrain 3DM-GX3 35	
Computer	Gumstix Verdex Pro	
GNSS Antenna	Maxtenna L1/L2 helical	
GNSS Receiver	Novatel OEMV1-df	
	Total Cost	**$17,900**

METHODS

Photogrammetric Reconstruction

The collinearity equations are the standard model used in photogrammetry [31] to relate 2D image projections to 3D space. The coordinates of an unknown object point A (i.e., snow surface) in a 3D Cartesian coordinate system can be related to a vector of image coordinates in the following manner. Note that equation 1 below is often augmented with terms that model the distortion of the camera lens, as described in [32]. However, for the sake of brevity they are left out.

$$\begin{pmatrix} x_{im} \\ y_{im} \end{pmatrix} = \begin{pmatrix} x_0 + f\dfrac{r_{11}(X_A - X_0) + r_{21}(Y_A - Y_0) + r_{31}(Z - Z_0)}{r_{13}(X_A - X_0) + r_{23}(Y_A - Y_0) + r_{33}(Z - Z_0)} \\ y_0 + f\dfrac{r_{12}(X_A - X_0) + r_{22}(Y_A - Y_0) + r_{32}(Z - Z_0)}{r_{13}(X_A - X_0) + r_{23}(Y_A - Y_0) + r_{33}(Z - Z_0)} \end{pmatrix} \quad (1)$$

where

x_{im}, y_{im}—measured image coordinates

f—focal length

X_o, Y_o, Z_o—Cartesian coordinates of camera perspective centre

X_A, Y_A, Z_A—Cartesian coordinates of object point A

r_{ij}—elements of rotation matrix R of angles (ω, ϕ, κ)

If only one image is available, then only the direction to the object point A can be determined, but not its absolute position in space. The 3D coordinates of A can only be computed if this spatial direction intersects another ray from at least one different image. The process of ray intersection and the simultaneous refinement of interior and exterior geometry according to an optimality criterion involving the image projections of all points is known as the bundle adjustment [31]. An excellent overview of bundle adjustment is given in [33].

The first stage of this photogrammetric workflow to reconstruct snow depth is to identify image projections of the same features in space from all of the camera views. For image matching, we used the Scale Invariant Feature Transform (SIFT), a technique developed by [34]. SIFT key features are identified in each image frame and then matched between frames using an approximate nearest neighbour kd-Tree approach [35]. After the key-points have been matched, an iterative Random Sample Consensus (RANSAC) approach following [36] is used to identify and eliminate outliers. After outlier removal, the image matches are stored for subsequent integration into the bundle adjustment.

Two bundle adjustment methodologies are used in the study. The first case employs a methodology similar to that of [20,21], in which GCP information is used to constrain the bundle adjustment; this is termed conventional aerotriangulation or indirect georeferencing. In the indirect case, the ground coordinates of a ground control point (X_A, Y_A, Z_A) are assigned high weights to constrain the bundle adjustment. The second case uses solely the perspective centre coordinates (X_o, Y_o, Z_o) obtained from the UAS navigation solution to constrain the weighted bundle adjustment. This method is similar to the one proposed in [28] and is also known as airborne aerotriangulation or alternatively, direct georeferencing. Because there is a spatial offset between the GPS antenna and the perspective centre of the camera, a lever arm offset was measured in the laboratory a priori and then applied post-flight. The a priori stochastic weighting criteria for airborne aerotriangulation is based on the uncertainty in the computed positions of the UAS platform using airborne GPS. The second method is more preferable than the first for obvious reasons, as no ground personnel are required to set ground control points within the measured scene. One-sigma values of 2 cm were assigned for the indirect case, as shown in Table 2. In both the direct and indirect BA, we assume that the interior orientation parameters are constant from frame to frame, known as the "block invariant" approach [37].

Table 2. The observations and their associated weights used in the bundle adjustment

Observation	Indirect (1σ)	Direct (1σ)
X_o, Y_o, Z_o (UAS GNSS)	1 m	0.03 m
X, Y, Z (GCP)	0.02 m	none
ω, ϕ, κ	2°	2°
x_o, y_o	3 pixels	3 pixels
f	3 pixels	2 pixels
k1, k2	0.1 pixels	0.1 pixels

Bundle adjustment is a non-linear process that requires reasonable initial approximation to ensure correct convergence. To generate initial approximations for scene structure, we use the well-known structure-from-motion (SfM) pipeline developed by [35], entitled Bundler. A comprehensive description of SfM algorithms can be found in [38]. The main input to Bundler is the set of image coordinates of the previously matched images. The output of the Bundler SfM algorithm is the reconstructed camera poses and a sparse set of scene structure, both of which are generated in an arbitrary coordinate system. To transform the camera positions and initial structure estimated using Bundler into a global coordinate system, we perform a seven-parameter Helmert transformation using the trajectory information provided by the UAS navigation solution. Once converted into a mapping coordinate system, we use the camera poses and structure as initial approximations into a final bundle adjustment. At this point, stochastic information that exists for different observations is included in the form of a covariance matrix to weight the adjustment. This ensures that the least squares solution converges to the most optimal values based on a priori error models. Should ground control observations be used to constrain the final adjustment, they are included in this step.

To properly integrate all available information about a priori error sources in the covariance matrix (or equivalently, the weight matrix) and generate a linearized approximation of the a posteriori covariance matrix, we utilize the Gauss-Helmert adjustment model, which is of the following form:

$$Bl + Ax = w \tag{2}$$

where A is the Jacobian matrix with respect to the parameters, and B is the Jacobian matrix with respect to the observations, and l and x are the observations and parameters, respectively. Assuming a known variance-covariance P for the observations, the weighted least squares solution would be the following:

$$x = (A^T (BP^{-1}B^T)^{-1} A)^{-1} * A^T (BP^{-1}B^T)^{-1} l \tag{3}$$

Or equivalently,

$$x = N^{-1} t \tag{4}$$

where

$$N = (A^T (BP^{-1}B^T)^{-1} A) \tag{5}$$

$$t = A^T (BP^{-1}B^T)^{-1} l \tag{6}$$

Because of the large amount of photos typically encountered in UAS photogrammetric applications, conventional matrix inversions are not optimal for solving the normal equations, due to the sparse nature of the A and B matrices. To improve performance and stability, the normal equations are partitioned into block form consisting of the camera and scene parameters [39], as shown below.

$$\begin{bmatrix} N_c & N_{cp} \\ N_{cp}^T & N_p \end{bmatrix} \begin{bmatrix} x_c \\ x_p \end{bmatrix} = \begin{bmatrix} t_c \\ t_p \end{bmatrix} \tag{7}$$

Note that self-calibration parameters (i.e., principle point, focal length, and lens distortion) can also be incorporated into this system of normal equations by extending the block structure above.

It should be noted that N_p is a block diagonal matrix, with blocks of size 3 × 3, and thus is trivial to invert for each block. Once N_p^{-1} has been obtained, we obtain a linear system consisting of just the camera parameters, as shown below.

$$S x_c = r \tag{8}$$

where

$$r = t_c - N_{cp} N_p^{-1} t_p \tag{9}$$

$$S = N_c - N_{cp} N_p^{-1} N_{cp}^{T} \tag{10}$$

The variable S, also known as the reduced camera matrix, is symmetric and positive-definite and can be solved using Cholesky factorization, after which the scene points can be estimated through back substitution in the following manner.

$$x_p = N_p^{-1}\,(t_p - N_{cp}^T x_c) \tag{11}$$

This method of solving the normal equations is known as the Schur complement trick [29,39]. The model iteratively converges on an exact solution that minimizes the sum of the squares of the residuals. The iteration process stops if a convergence criterion is reached, or if the number of iterations exceeds a user-defined maximum. We performed the bundle adjustment in a projected coordinate system, specifically Universal Transverse Mercator Zone 55S. After the bundle adjustment has been performed, the final stage of this workflow is to perform a dense multi-view stereo reconstruction of the aligned images to generate the dense 3D scene geometry. Many different approaches can be taken, and an excellent review of multi-view stereo techniques can be found in Seitz et al. [40]. For this task, we utilize an existing multi-view stereo reconstruction routines, specifically Patch-based Multiview Stereo (PMVS2), developed by Furukawa & Ponce [41]. The algorithm starts by detecting features in each image, matches them across multiple images to form an initial set of patches, after which visibility constraints are applied to filter away false matches. The patch model is then converted to mesh, after which a refinement algorithm is used to achieve higher accuracy. As snow is a relatively homogeneous surface, we modified PMVS in several ways for better performance. This included raising the normalized cross-correlation threshold for matching to 0.8, as well as enlarging the kernel size that is used to compute the photometric consistency score. We were fortunate that most of our images were well-illuminated, sharp, and not overexposed. It should be noted that the image quality had a direct effect on the reconstruction precision, and sub-optimal imagery was discarded a priori.

The output of PMVS is a dense point cloud of the estimated scene structure that can then be used for subsequent surface computations. The point clouds are generally quite dense (up to 200–400 points/m^2), though certain areas were much sparser due to camera exposure, scene illumination (shadows), and image blur. We used an averaging method to compute the height coordinate at the location of the depth measurement of the surface for differencing. All points falling within a 5 cm radius of the GPS snow depth observation were collected, from which the mean height was computed. This was done for both snow-free and snow-covered scenes. Figure 4 shows the overall work flow employed in this study.

To ensure that our techniques and workflow are in line with the current state of the art, we compared our results to those obtained using Agisoft Photoscan, a commercial software that is used extensively within the UAS scientific community. We sampled both dense point clouds at 5 cm resolution using nearest neighbour interpolation and computed the resulting difference in Z. As shown in Figure 5, our results are very similar to those obtained using Agisoft. The mean of the difference was −0.01 m, and the standard deviation was 0.04 m, after outliers are removed.

Error Sources: Snow and Vegetation

The vegetation in the study area consists of eucalyptus and spruce trees, along with considerable dense shrubbery ranging from 0.3–1.2 m in height (e.g., Figure 1). During the snow-covered data collection, the GPS observations of snow depth were made without a priori knowledge of underlying vegetation. When making many of the depth measurements, the survey pole was inadvertently placed directly within the middle of shrubs lying below the snow surface. Thus the in situ depth measurements were measuring the combined snow depth and shrub height. However, the photos and photogrammetry techniques will effectively be measuring the depth of snow above the shrub layer. Such in situ measurements should not be used to evaluate photogrammetric depth estimate performance.

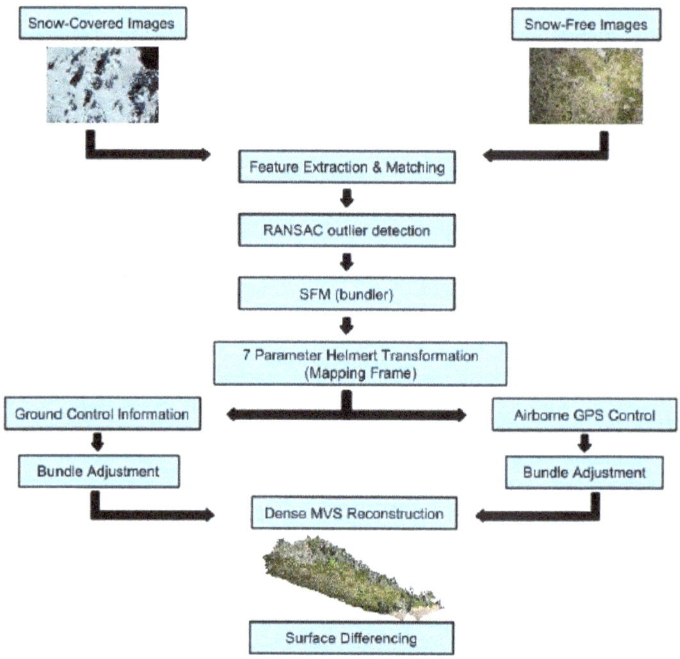

Figure 4. The overall workflow of the study. Note that two different bundle adjustment strategies are implemented; one involving conventional methods (**left**); the other using airborne aero-triangulation (**right**).

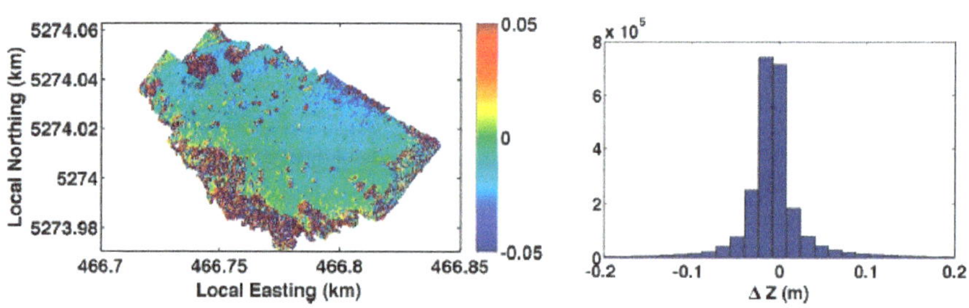

Figure 5. Comparison of our workflow with the current commercial state of the art (**left**) using differenced DEMs. Histogram of the Z differences (**right**).

We used two methods to reduce the effect of vegetation on the retrieved snow depths. First, we eliminated observations of snow depth that were taken in the midst of thick stands of shrubs, based on visual analysis of the point cloud. This was implemented due to the higher likelihood of incorrect measurements of in situ measured GPS depths. While the removal of measurements is not an ideal scenario, it is nonetheless necessary because there is no guarantee of accuracy on the "truth" in situ observations. This reduced the overall number of in situ measured snow observations from 37 to 20.

Secondly, we filtered the raw point cloud of in situ measurements of the snow-free and snow-covered datasets to create a bare earth point cloud model. To filter the vegetation from the point cloud and create a bare earth model, we used the Multi Scale Curvature Classification (MCC) filtering technique [42]. MCC uses a thin plate spline to filter point clouds based on an iterative multi-scale assessment of non-ground curvature. While the filtering was effective at removing larger trees, small shrubs and grass patches are non-trivial to remove because they have very similar characteristics to the overall terrain variability.

While our study area required filtering of the point cloud for bare earth calculation, this may not be required for areas that have little to no vegetation. Such study sites are desirable for snow depth retrieval based on surface differencing, but are not always available in practice.

RESULTS

Predicted Snow Depth Accuracy

Snow depths determined photogrammetrically are computed by differencing two or more surface datasets. As each dataset is calculated independently, the accuracy of the estimated snow depth is thus a function of the accuracy of the datasets from which it is computed. Assuming the snow-free (SF) and snow-covered (SC) surfaces have an unbiased, normally distributed measurement error, the predicted error in snow depth can be estimated from the overall error of the two surface datasets, σ^2_{SF}, σ^2_{SC}, in the following manner.

$$\sigma_{PE} = \sqrt{\sigma_{SF}^2 + \sigma_{SC}^2} \tag{12}$$

To estimate the error of the different photogrammetric bundle adjustments used in this study (σ^2_{SF}, σ^2_{SC}), we use a jackknife validation [43]. Five measured ground control points were left out of a conventional BA constrained by ground control points, and the GPS field-measured coordinates were compared to those estimated from the 3D point cloud that resulted from the reconstruction process, as shown in Figure 6. The overall error estimate is calculated by calculating the RMSE based on ΔX, ΔY, and ΔZ. The errors for the two surface datasets reconstructed using conventional BA are 6.2 cm and 5.9 m, respectively. Using Equation (4), the predicted error (σ_{PE}) for the snow depth reconstruction is 8.6 cm.

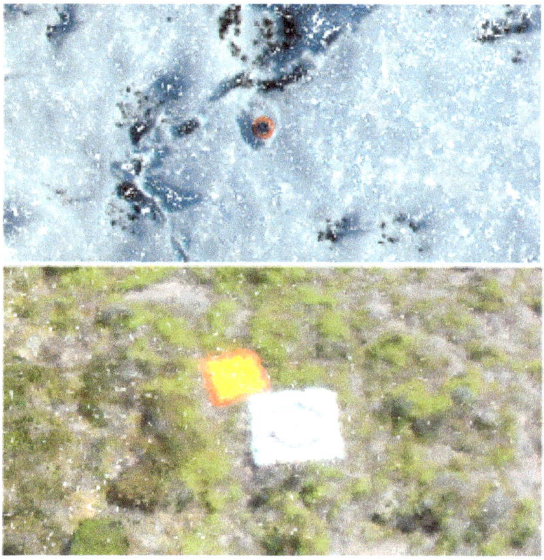

Figure 6. Bootstrap validation was carried out by finding the photo identifiable control points in the reconstructed point cloud (shown above), and comparing the estimated coordinates with those measured from the RTK GPS.

Snow Depth via Conventional BA with GCPs

The snow-free and snow-covered datasets in the conventional case correspond to bundle adjustments that incorporated ground control points (GCPs) to constrain the 3D reconstruction. Figure 7 shows the in situ observed snow depth plotted against the photogrammetrically reconstructed snow depth estimated using the conventional BA workflow (Section 3). The root mean square error (RMSE) of the conventional BA derived snow depths is 9.6 cm, with a bias of 3.4 cm. The true RMSE is 1 cm (12%) larger than the predicted error in the two-point cloud datasets calculated via Equation (4). This is likely due to residual uncertainties due to the vegetation that characterizes our study area.

Additionally, the predicted error computed in Equation (4) did not include the measurement errors uncertainty associated with the in situ snow depths calculated from differenced GPS observations. This is an incorrect assumption, as these observations are also characterized by random errors. If we assume the GPS observations have a 1σ of 2–4 cm, then the overall predicted error is of the order of 10 cm, which is closer to the RMSE that was observed. One interesting statistic to note is the relative error of the estimates as a function of snow depth. At the lower depths (e.g., 35 cm), the relative error is 27.6% of total snow depth. However, at higher snow depths (e.g., 120 cm) the relative error only amounts to 8% of the total snow depth. Thus it is reasonable to assume this technique is advantageous for areas characterized by steady snowfall accumulation and high snow depth.

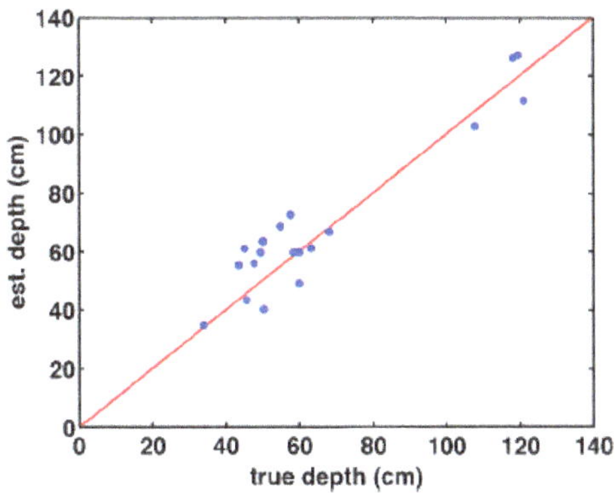

Figure 7. Scatter plot of observed snow depth vs. estimated snow depth calculated from conventional BA techniques and GCPs (n = 20), with an RMSE of 9.59 cm and R^2 value of 0.91.

The correlation coefficient (r) between the GPS observed in situ snow depth observations and the photogrammetrically estimated snow depths from the conventional BA workflow was 0.95. This means the photogrammetrically resolved snow depths explain 91% (r^2) of the variance in the in situ observed GPS snow depths.

Snow Depth via Direct BA Using Airborne GPS

The snow-covered dataset in the direct case corresponds to a bundle adjustment which incorporated no GCPs to constrain the 3D re-construction; only trajectory information from the kinematic airborne GPS observations was used. Figure 8 shows the GPS-observed snow depth plotted against the photogrammetrically reconstructed snow depth estimated using the direct BA workflow described in Section 3. There is a clear negative bias in the error associated with snow depth retrieval using the GPS-based aerial triangulation methods, with a RMSE of 18.4 cm. The relative error of the snow depth estimates determined via direct BA with bias, ranged from 52.7% to 15% at depths of 35 and 120 cm, respectively.

The reason for this bias is unknown. Initially, it was assumed that the error was in the estimate of focal length calculated as part of the adjustment. However, when the focal length was constrained to the same calibrated values determined in the GCP-based adjustment described in Section 4.1, the vertical bias still remained. Theoretically, this problem should not exist if proper care is taken to ensure that the correct stochastic information associated with the observations and parameters is incorporated into the bundle adjustment. Given that the integer ambiguities have been correctly resolved in the GPS solution process, and that the lever arm offset has also been completed, airborne GPS-based AT should provide sufficient accuracy without the addition of ground control features [23]. In practice however, the vertical bias exhibited in our data is commonly encountered within GPS-based AT projects. The most common way to resolve this is by introducing a minimal number of ground control points into the BA solution. A description of such a process is found in [31]. Further analysis into why this systematic error exists remains a future area of research.

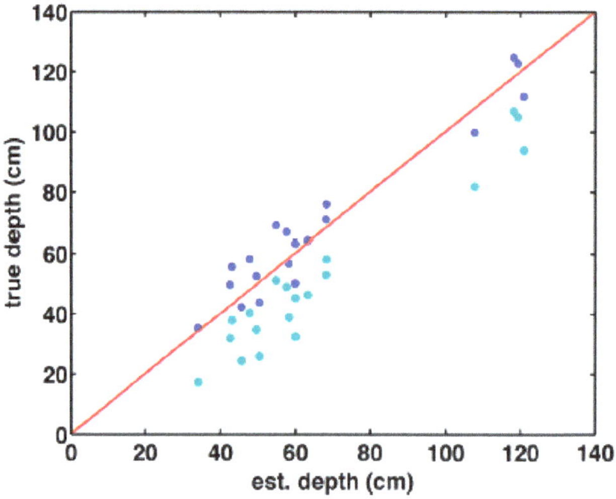

Figure 8. Scatter plot of the observed snow depth vs. estimated snow depth calculated using the direct BA techniques. The cyan dots represent the original estimated snow depths (with bias), whereas the blue dots represent snow estimates when the affine transformation is applied. The RMSE was 18.4 cm with using only onboard GPS; after the transformation was applied, the RMSE fell to 8.8 cm.

Co-Registration Using Affine Transformation

For estimation of snow depth in many areas, it turns out that the bias introduced by direct BA is of no consequence. This is because we are only interested in relative changes between the snow-free and snow-covered surfaces. Thus, if a fixed point/s in space is visible in both datasets, they can be co-registered by means of an affine transformation [44]. This is done by determining the 3D difference of the fixed surface point/s, and adding a fixed translation (the difference) to each point in the corresponding dataset. The implied assumption is that the two surface datasets are conformal and do not have warping.

In our experiment, we used a single fixed point on the corner of a large boulder on the southern end of our test sight, as shown in Figure 9. The difference of the boulder corner is determined between the snow-covered and snow-free point clouds, after which the translation is applied to co-register the two surfaces. After co-registration, the overall RMSE of the reconstructed snow depth is 8.8 cm (Figure 7), with an r^2 value of 0.96. Such results are attainable because relative accuracy of photogrammetrically derived surfaces is very high due to the re-projection cost function that is minimized in the bundle adjustment. Thus, while absolute errors such as those described in Section 4.3 may be difficult to mitigate, computations based on the relative relationships between the two surfaces can be quickly and accurately made.

Figure 9. For co-registering the two datasets, we used the corner of a large boulder (red star) within the study site to estimate the translational components for the affine transformation.

High-Resolution Snow Depth Estimates

The spatially continuous snow depths over our study area are shown in Figure 10. The spatial variation in snow depth is readily apparent in photogrammetrically estimated snow depths in Figure 10, while the overall trends are consistent when compared to Figure 3. The mean snow depth from our study area according to the in situ measured GPS observations was 61.8 cm, with a standard deviation of 14.3 cm. The mean value of the spatially continuous snow depths within our study area computed from the indirect BA workflow is 61.9 cm. While this minor difference is somewhat unrealistic given the overall magnitude of the error budget, it nonetheless provides additional methodological validation.

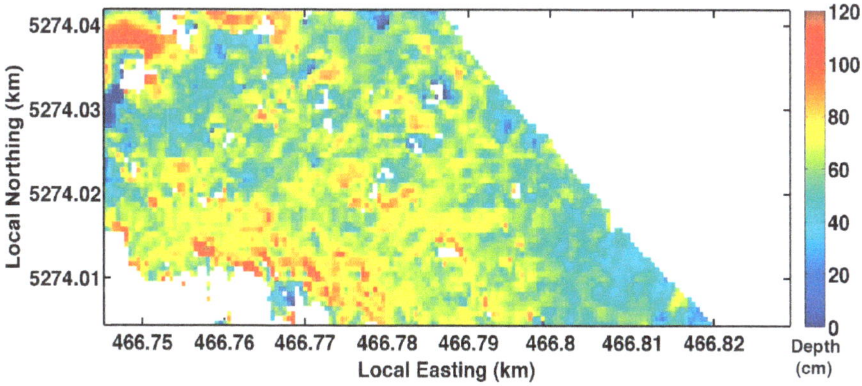

Figure 10. High-resolution (50 cm) snow depths over our study area, as computed from the differenced point cloud measurements.

DISCUSSION AND CONCLUSIONS

Discussion

The experiences gained so far from extracting snow depth from optical imagery via BA show that the two methodologies outlined herein have different strengths and weaknesses for the various data processing tasks. Conventional BA routines with ground control points provide accurate resolution of snow depths, with an overall RMSE of 9.6 cm. Direct BA mitigates the need for obtaining ground control points (GCPs) to constrain the adjustment, but the overall accuracy (RMSE) is 18.4 cm. However, if a fixed surface point is co-visible in both the snow-free and snow-covered datasets, the bias can be removed and the point clouds aligned via an affine transformation. In our case, this effectively increased the accuracy to 8.8 cm RMSE. If co-registration is possible, we hypothesize that navigation solutions based on psuedorange measurement alone (~3 m accuracy) could be used to successfully measure spatially continuous snow depths, thus further lowering the cost associated with this type of measurement. Both direct and indirect reconstructions exceeded the snow depth accuracy, precision, and resolution of prior efforts [19–21].

While the spatial scale of our study site is relatively small, the techniques and methods described in this study are well-suited for integration at larger scales with platforms that have longer flight time capabilities, such as fixed-wing UAS. The overall workflow is relatively straightforward, and few physical assumptions need to be made in the retrieval process, in contrast to e.g., passive microwave-based retrievals (e.g., [45,46]). This simple, straightforward approach is advantageous for government agencies looking for accurate, high-resolution measurement techniques that can be used at a moment's notice. It should be mentioned that snow depth estimates made via photogrammetric surface differencing cannot not automatically be converted to the snow water equivalent (SWE) amount that most forecasting agencies desire, as this requires knowledge of snow density. However, recent literature has suggested it is possible to characterize SWE based on depth measurements alone, due to the fact that SWE is more closely linked to depth than it is to density [47].

In our study area, vegetation has adverse effects on the accuracy of snow depth measurements made using photogrammetric techniques because of its correlation with terrain variability. One way to potentially mitigate the vegetation problem encountered in surface differencing approaches such as this is to augment the photogrammetric observations with co-registered LiDAR measurements. With the miniaturization of LiDAR sensors in recent years [17], a combined UAS-based LiDAR/photogrammetric system could provide ground returns in areas obscured by vegetation to the photogrammetric reconstruction, along with providing general redundancy to the overall surface dataset. This is an exciting possibility and should be investigated in future studies.

Additionally, the photogrammetric method described herein can be combined with existing remote sensing platforms and retrieval approaches to enhance snow property estimates at larger scales using data assimilation techniques. Data assimilation optimally combines the large spatial coverage of spaceborne measurements, physically-based land surface models for prediction, and the high accuracy and resolution of UAS snow depth retrievals to generate highest likelihood estimates of snow states based on the stochastic properties of each input. This is the current state of the art in snow forecasting, and is used globally with encouraging results [9]. Future studies should examine the feasibility of integrating snow depth estimates obtained from photogrammetric measurements on-board UAS platforms into such largescale retrieval algorithms.

While our results indicate that this method holds great promise, we believe there is still considerable improvement to be made. For discussion sake, we have identified three areas which we believe deserve further exploration. The first is understanding how the physical state of the snow itself effects the reconstruction quality. Does older metamorphosed snow (sublimation, melt/refreeze, etc.) provide more accurate results than freshly fallen flakes, because of the increased spectral variability? In the case of this study, the snow was quite fresh (2 days) but had significant water content and was beginning to melt, which may have been beneficial to this study. The second area is gaining a better understanding of how the overall accuracy of the snow surface reconstruction is affected by the acquisition process. For example, do sunny days provide the best results, or is the overall accuracy invariant to weather conditions? What will be the trade-off between snow depth accuracy and flying height? In general, these concepts are well understood within the photogrammetric community, but perhaps have yet to be applied to the specific application of snow depth retrieval. The third and final area is simply improvement and innovation in the workflow. Custom algorithms specifically tailored to this application are certain to perform with higher reliability as compared to generalized techniques. This is why we believe it is in the community's best interest to become less dependent on commercial software offerings. In that respect, this proposed workflow provides an excellent starting point for continued exploration and innovation.

CONCLUSION

A novel technique has been described to measure high-resolution spatial distributions of snow depths accurately and affordably. While it will not provide the global spatial and temporal coverage of existing airborne and spaceborne sensors (e.g., passive microwave), it can nonetheless provide reliable information about snow processes at local measurement scales. This represents an important contribution to the snow science community, as accurate representations of snow distribution are needed for improvements to hydrological forecasts, climate models, and for the future testing and validation of remote-sensing retrieval algorithms. Smaller-scale process-oriented snow studies could benefit from such observations, especially as difficulty persists in existing methods to create spatially-continuous snow depth maps from point observations [5].

We successfully estimated snow depth at high spatial resolution in mountainous terrain with thick vegetation cover based on photogrammetrically derived 3D surface models from aerial photography acquired by a micro unmanned aircraft system (UAS). Snow depths were estimated using conventional BA techniques with an overall RMSE of 9.6 cm, whereas direct BA techniques without GCP information yielded an RMSE of 18.4 cm. If an affine transformation is applied using a fixed surface point that is co-visible in both the snow-free and snow-covered datasets, the RMSE is reduced to 8.4 cm. This is the most convenient way to resolve systematic error that may be present; however, it is dependent on having points on both surfaces that can be co-registered. Physical characteristics of this environment (e.g., variable terrain, deep snow, etc.) which can be highly problematic for other remote sensing based snow retrieval methods do not significantly affect the accuracy of the estimated snow depth, in our case.

The research results demonstrate that airborne remote sensing using UAS platforms which implement low-cost COTS digital camera systems and standard photogrammetric techniques can provide snow depth retrieval information at a level of accuracy (<10 cm) that can effectively support those in the snow community which have a need for high resolution measurements of snow depth at local measurement scales. This technology will only continue to improve in reliability, accuracy, and spatial scalability. The proliferation of UAS technology and improved sensor capabilities should provide for rapid advances in abilities to retrieve snow properties in mountainous areas which are often characterized by high snowfall amounts, but dangerous and remote measurement requirements.

ACKNOWLEDGMENTS

This project was funded through a National Science Foundation East Asia and Pacific Studies Institute (EAPSI) fellowship, as well as National Aeronautics and Space Administration NESSF fellowship grant NNX11AL41H. Additionally, the authors acknowledge the Winfred Violet Scott Trust for funding support in purchasing the OktoKopter that was used in this study. Helge Aasen and Juliane Bendig are acknowledged for their help with the GPS survey during the February 2014 survey.

REFERENCES

1. 1. Barnett, T.P.; Adam, J.C.; Lettenmaier, D.P. Potential impacts of a warming climate on water availability in snow-dominated regions. Nature **2005**, 438, 303–309.
2. Hu, J.I.A.; Moore, D.J.; Burns, S.P.; Monson, R.K. Longer growing seasons lead to less carbon sequestration by a subalpine forest. Glob. Chang. Biol. **2010**, 16, 771–783.
3. Peck, E.L. Snow measurement predicament. Water Resour. Res. **1972**, 8, 244–248.
4. Hall, D.K.; Sturm, M.; Benson, C.S.; Chang, A.T.C.; Foster, J.L.; Garbeil, H.; Chacho, E. Passive microwave remote andin situ measurements of artic and subarctic snow covers in Alaska. Remote Sens. Environ. **1991**, 38, 161–172.
5. Erxleben, J.; Elder, K.; Davis, R. Comparison of spatial interpolation methods for estimating snow distribution in the Colorado Rocky Mountains. Hydrol. Process. **2002**, 16, 3627–3649.
6. Chang, A.T.C.; Foster, J.L.; Hall, D.K. Nimbus-7 SMMR derived global snow cover parameters. Ann. Glaciol. **1987**, 9, 39–44.

7. Foster, J.L.; Hall, D.K.; Chang, A.T.C.; Rango, A. An overview of passive microwave snow research and results. Rev. Geophys. **1984**, 22, 195–208.
8. Foster, J.L.; Chang, A.T.C.; Hall, D.K. Comparison of snow mass estimates from a prototype passive microwave snow algorithm, a revised algorithm and a snow depth climatology. Remote Sens. Environ. **1997**, 62, 132–142.
9. Takala, M.; Luojus, K.; Pulliainen, J.; Derksen, C.; Lemmetyinen, J.; Kärnä, J.P.; Bojkov, B. Estimating northern hemisphere snow water equivalent for climate research through assimilation of space-borne radiometer data and ground-based measurements. Remote Sens. Environ. **2011**, 115, 3517–3529.
10. Vander Jagt, B.J.; Durand, M.T.; Margulis, S.A.; Kim, E.J.; Molotch, N.P. The effect of spatial variability on the sensitivity of passive microwave measurements to snow water equivalent. Remote Sens. Environ. **2013**, 136, 163–179.
11. Deems, J.S.; Painter, T.H.; Finnegan, D.C. Lidar measurement of snow depth: A review. J. Glaciol. **2013**, 59, 467–479.
12. Colomina, I.; Molina, P. Unmanned aerial systems for photogrammetry and remote sensing: A review. ISPRS J. Photogramm. Remote Sens. **2014**, 92, 79–97.
13. Eisenbeiss, H.; Zhang, L. Comparison of DSMs generated from mini UAS imagery and terrestrial laser scanner in a cultural heritage application. Int. Arch. Photogramm. Remote Sens. Spat. Inf. Sci. **2006**, 36, 90–96.
14. Jaakkola, A.; Hyyppä, J.; Kukko, A.; Yu, X.; Kaartinen, H.; Lehtomäki, M.; Lin, Y. A low-cost multi-sensoral mobile mapping system and its feasibility for tree measurements. ISPRS J. Photogramm. Remote Sens. **2010**, 65, 514–522.
15. Lucieer, A.; de Jong, S.; Turner, D. Mapping landslide displacements using Structure from Motion (SfM) and image correlation of multi-temporal UAS photography. Progress Phys. Geogr. **2013**.
16. Ryan, J.C.; Hubbard, A.L.; Todd, J.; Carr, J.R.; Box, J.E.; Christoffersen, P.; Snooke, N. Repeat UAV photogrammetry to assess calving front dynamics at a large outlet glacier draining the Greenland Ice Sheet. Cryosphere Discuss. **2014**, 8, 2243–2275.
17. Wallace, L.; Lucieer, A.; Watson, C.; Turner, D. Development of a UAS-LiDAR system with application to forest inventory. Remote Sens. **2012**, 4, 1519–1543.
18. Harwin, S.; Lucieer, A. Assessing the accuracy of georeferenced point clouds produced via multi-view stereopsis from unmanned aerial vehicle (UAS) imagery. Remote Sens. **2012**, 4, 1573–1599.
19. Tao, W.; Lei, Y.; Mooney, P. Dense point cloud extraction from UAV captured images in forest area. In Proceedings of the 2011 IEEE International Conference on Spatial Data Mining and Geographical Knowledge Services (ICSDM), Fuzhou, China, 29 June–1 July 2011; pp. 389–392.
20. Cooper, C.F. Snow cover measurement. Photogramm. Eng. **1965**, 31, 611–619.
21. Smith, F.M.; Cooper, C.F.; Chapman, E.G. Measuring Snow Depths by Aerial Photography. In Proceedings of the 35th Annual Western Snow Conference, Boise, ID, USA, 18–20 April 1967; pp. 66–72.

22. Nolan, M.; Larsen, C.F.; Sturm, M. Mapping snow-depth from manned-aircraft on landscape scales at centimeter resolution using Structure-from-Motion photogrammetry. Cryosphere Discuss. **2015**, 9, 333–381.
23. Ledwith, M.; Lundén, B. Digital photogrammetry for air-photo-based construction of a digital elevation model over snow-covered areas-Blamannsisen, Norway. Nor. J. Geogr. **2001**, 55, 267–273.
24. Ebadi, H. A Comprehensive Study on GPS-Assisted Aerial Triangulation. Ph.D. Dissertation, University of Calgary, Calgary, AB, Canada, 1997.
25. Cramer, M.; Stallmann, D.; Haala, N. Direct georeferencing using GPS/inertial exterior orientations for photogrammetric applications. Int. Arch. Photogramm. Remote Sens. Spat. Inf. Sci. **2000**, 33, 198–205.
26. Mostafa, M.M.; Schwarz, K.P. Digital image georeferencing from a multiple camera system by GPS/INS. ISPRS J. Photogramm. Remote Sens. **2001**, 56, 1–12.
27. Lee, C.Y.; Jones, S.D.; Bellman, C.J.; Buxton, L. DEM creation of a snow covered surface using digital aerial photography. Int. Arch. Photogramm. Remote Sens. Spat. Inf. Sci. **2008**, 37, 18–27.
28. Lemmon, T.R.; Gerdan, G.P. The Influence of the Number of Satellites on the Accuracy of RTK GPS Positions. Aust. Surv. **1990**, 44, 64–70.
29. Turner, D.; Lucieer, A.; Wallace, L. Direct georeferencing of ultrahigh-resolution UAV imagery. IEEE Trans. Geosci. Remote Sens. **2014**, 52, 2738–2745.
30. Van der Merwe, R.; Wan, E.A. Sigma-point Kalman filters for integrated navigation. In Proceedings of the 60th Annual Meeting of the Institute of Navigation, Dayton, OH, USA, 7–9 June 2004; pp. 641–654.
31. Mikhail, E.M.; Bethel, J.S.; McGlone, J.C. Introduction to Modern Photogrammetry; John Wiley & Sons Inc.: New York, NY, USA, 2001; Volume 1.
32. Brown, D. Close-range camera calibration. Photogramm. Eng. **1971**, 37, 855–866.
33. Triggs, B.; McLauchlan, P.F.; Hartley, R.I.; Fitzgibbon, A.W. Bundle adjustment—A modern synthesis. In Vision Algorithms: Theory and Practice; Springer: Berlin/Heidelberg, Germany, 2000; pp. 298–372.
34. Zhang, L.; Gruen, A. Multi-image matching for DSM generation from IKONOS imagery. ISPRS J. Photogramm. Remote Sens. **2006**, 60, 195–211.
35. Scambos, T.A.; Dutkiewicz, M.J.; Wilson, J.C.; Bindschadler, R.A. Application of image cross-correlation to the measurement of glacier velocity using satellite image data. Remote Sens. Environ. **1992**, 42, 177–186.
36. Lowe, D.G. Object recognition from local scale-invariant features. In Proceedings of the Seventh IEEE International Conference on Computer Vision, Kerkyra, Greece, 20–27 September 1999; Volume 2, pp. 1150–1157.
37. Snavely, N.; Seitz, S.M.; Szeliski, R. Photo tourism: Exploring photo collections in 3D. ACM Trans. Graph. **2006**, 25, 835–846.

38. Nistér, D. Preemptive RANSAC for live structure and motion estimation. Mach. Vis. Appl. **2005**, 16, 321–329.

39. Brown, D.C. The bundle adjustment—Progress and prospects. Int. Arch. Photogramm. Remote Sens. Spat. Inf. Sci. **1976**,21, 1.

40. Huang, T.S.; Netravali, A.N. Motion and structure from feature correspondences: A review. Proc. IEEE **2004**, 82, 252–268.

41. Agarwal, S.; Snavely, N.; Seitz, S.M.; Szeliski, R. Bundle adjustment in the large. In Computer Vision–ECCV 2010; Springer: Berlin/Heidelberg, Germany, 2010; pp. 29–42.

42. Seitz, S.M.; Curless, B.; Diebel, J.; Scharstein, D.; Szeliski, R. A comparison and evaluation of multi-view stereo reconstruction algorithms. In Proceedings of the IEEE Computer Society Conference on Computer Vision and Pattern Recognition, Rio de Janeiro, Brazil, 17–22 June 2006; Volume 1, pp. 519–528.

43. Furukawa, Y.; Ponce, J. Accurate, dense, and robust multiview stereopsis. IEEE Trans. Pattern Anal. Mach. Intell. **2010**,32, 1362–1376.

44. Evans, J.S.; Hudak, A.T. A multiscale curvature algorithm for classifying discrete return Lidar in forested environments. IEEE Trans. Geosci. Remote Sens. **2007**, 45, 1029–1038.

45. Efron, B.; Gong, G. A leisurely look at the bootstrap, the jackknife, and cross-validation. Am. Stat. **1983**, 37, 36–48.

46. Howat, I.M.; Smith, B.E.; Joughin, I.; Scambos, T.A. Rates of southeast Greenland ice volume loss from combined ICESat and ASTER observations. Geophys. Res. Lett. **2008**, 35, L17505.

47. Durand, M.; Kim, E.J.; Margulis, S.A. Radiance assimilation shows promise for snowpack characterization. Geophys. Res. Lett. **2009**, 36.

48. Sturm, M.; Taras, B.; Liston, G.E.; Derksen, C.; Jonas, T.; Lea, J. Estimating snow water equivalent using snow depth data and climate classes. J. Hydrometeorol. **2010**, 11, 1380–1394.

3
CHAPTER

CHANGE DETECTION OF DESERT SAND DUNES: A REMOTE SENSING APPROACH

Surender Varma, Vatsal Shah, Biplab Banerjee and Krishna Mohan Buddhiraju

Keywords: Change Detection; Desert Land Forms; Monitoring Sand Dunes

INTRODUCTION

Desert regions are increasingly facing imminent threats and exceptionally vulnerable to effects of climate changes, urbanization, land degradation, water shortage, etc. [1]. Deserts are extremely important habitats for specialized flora and fauna. In addition, deserts possess extensive amounts of minerals, such as lithium and boron. Low surface runoff, one of the major characteristics of desert regions, helps in preserving these minerals within the land. Quantification of dune morphology not only helps for mapping purposes, but also provides important information about the regional geologic history, environmental change, and desertification. Knowledge about changes occurring in dune boundaries and their directions will also contribute to the ability in predicting wind patterns. From a human perspective, if the dunes are changing in direction towards populated regions, immediate precautionary steps need to be initiated by the respective authorities. Hence, it has become imperative to accurately map and monitor deserts to formulate and implement various management activities. Categorizing dunes using field instruments is generally arduous as they generate sparse data. Quantitative methods using remote sensing are more viable for data collection and subsequent monitoring of dunes.

Major studies on desert dynamics are mainly focused on changes that occur in and around vegetation or water bodies. Dunes can either be stable or mobile, depending on vegetation and wind power. Unrestrained erosion is generally a feature of bare dunes.

Citation: Varma S., Shah V., Banerjee B. and Buddhiraju K.M. (2014) Change Detection of Desert Sand Dunes: A Remote Sensing Approach. http://dx.doi.org/10.4236/ars.2014.31002.

Stabilized dunes, on the other hand, do not undergo major changes in texture, and can thus facilitate the growth of vegetation [2] [3]. In a study by [4], a combination of Digital Elevation Modelling (DEMs)` and GPS guided field surveying was used to obtain base data on dunes. The NDVI index is used frequently to monitor the growth of vegetation and its relation to surrounding settlements. [5] used a brightness index, semi-variograms, and other statistical data to analyse the changes in texture orientations, while [6] applied Principal Component Analysis (PCA) to detect changes within images of desert cities.

While the above techniques are able to capture certain aspects of desert environments, they are not specifically devised towards the monitoring of the orientation of dunes. Analysing desert areas using satellite imagery is a challenging task, due to weak textural differences and the close to homogeneous spectral responses in most parts of the terrain. Hence the categorization of dunes via traditional clustering techniques is a challenging task, as dunes with differing orientations will have similar grey levels, as well as NDVI and brightness index values.

Researchers in the remote sensing community have extensively worked on a variety of methodologies for other land cover types, but few methodologies have been proposed specifically for dune regions. In this paper we present a novel way of identifying changes in dune orientations.

Radon Transform

Radon developed the Radon transform in the early part of 19th century [7]. A line with normal vector in the direction θ at a distance of s from the origin satisfies the line equation (1):

$$x\cos\theta + y\sin\theta - s = 0 \tag{1}$$

$$g(s,\theta) = \iint f(x,y)\cdot\delta(x\cos\theta + y\sin\theta - s)\,dxdy \tag{2}$$

The mass density $f(x, y)$ evaluated over different viewing directions will give rise to a series of 1D projections parameterized by s. The Dirac delta function δ, is zero for every argument apart from those on the s axis. Its integral, $g(s,\theta)$, is the radon transform, as given in Equation (2). The maximum angle of the radon spectrum can be used to represent surface orientations of sand dunes in [8]. The radon spectrum also contains additional information can be captured using statistical features that can be used to delineate dunes. In a related study by [9], the Radon transform was used to compute the global orientation of the texture present in the image. In this study, statistical features computed from the radon spectrum are initially used to cluster various texture regions. A change detection technique is subsequently proposed to monitor the changes occurring in these dune texture boundaries.

The study areas selected belong to various dune regions worldwide. Depending on the availability of good quality data, it has been ensured that the images selected correspond to the same time of the year. This is a prerequisite for change detection to ensure that the changes occurring are not due to seasonal changes.

METHODOLOGY

The main objective of this methodology is to identify regions of change in the multi-temporal desert dune datasets. A post-clustering based approach was developed for identifying changes in the orientation of dunes. Initially, contiguous regions representing dunes, with each region having similar dune orientation, are generated. The region boundaries are used as a basis for identifying changes in multi-temporal remote sensing data on a temporal scale. If the boundaries of the corresponding regions match, it is considered that there is "no-change" in dune patterns, otherwise it is considered that "change" has occurred.

The methodology adopted is shown in the flowchart of Figure 1. The input image and the reference image belonging to different times, but corresponding to the same region. are first registered to ensure the correctness of corresponding pixels for the change detection algorithm. Histogram matching is subsequently performed to ensure that the changes observed are not due to radiometric differences. Radon transform is then applied on the imagery, and similar to the Haralick texture features [10], three statistical features (Equations (3), (4), and (5)) based on the radon spectrum are calculated: the mode, standard deviation, and entropy. These features are seen to be superior over the mean, median and correlation features for dune texture delineation. These statistical quantities are extracted at each pixel location using a moving window (block) processing technique:

1) Mode radon angle

$$\theta_{mode} = \{\theta_{max} : f(\theta_{max}) \geq f(\theta), \forall \theta \in [0, \pi]\} \tag{3}$$

2) Entropy

$$\theta_{entropy} = -\sum_{i=0}^{\pi} p(\theta_i)\ln[p(\theta_i)] \text{ where } p(\theta_i) = \frac{\theta_i f(\theta_i)}{\sum_{i=0}^{\pi} f(\theta_i)} \tag{4}$$

3) Standard deviation

$$\theta_{sd} = \sqrt{\frac{\sum_{i=0}^{\pi} f(\theta_i)[\theta_{mean} - \theta_i]^2}{\sum_{i=0}^{\pi} f(\theta_i)}} \tag{5}$$

In Equations (3) to (5), the function $f(\theta)$ denotes the number of pixels with angle θ within the window centred around the pixel in consideration. The mode angle given in Equation (3) is the angle with the highest frequency within the window. Entropy represents the randomness of the angular information. In Equation (4), $p(\theta)$ is the fraction of pixels having orientation θ within the neighbourhood. The standard deviation relates to the spread of angles with respect to the mean angle.

Three different images with the same size as the original image are created using the above calculated statistical features for each block.

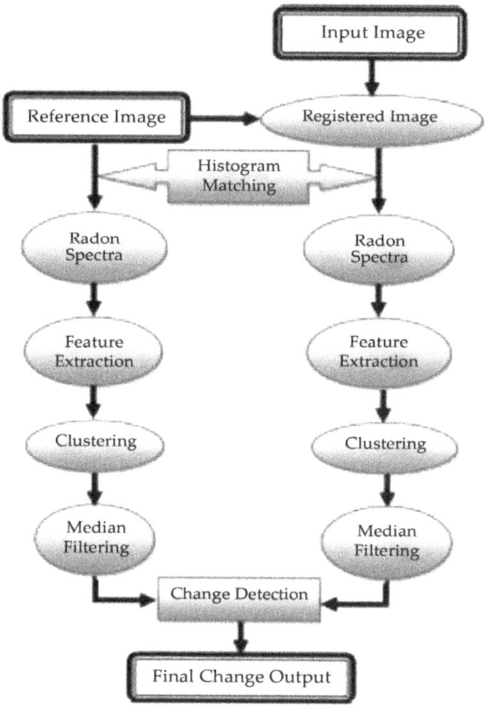

Figure 1: Flowchart of the methodology.

Next, each pixel is assigned the same attributes as that of the majority of pixels within a block. Thresholds are assigned empirically using trial and error, based on their ability to delineate dune patterns. A binary image is thus created for each feature. Once again, these discrete binary feature outputs are transformed into continuous feature output images using a recursive smoothing step. Finally, the texture clustered output is obtained by combining the feature outputs using the logical "or" operator.

The window based processing technique is used in order to get contiguous cluster regions. Window size is dependent on a host of factors, such as the image resolution, the size of the texture patterns on the dunes and the distance between successive dune crests. The window size, the thresholds used in clustering and other such parameters used in the clustering algorithm are kept constant for both the reference and the input images.

The clustered regions obtained from the reference and the registered input images are then compared with each other. The changes occurring in the dune orientations are captured by observing the variations in the boundaries of the clustered regions for identical dune orientations. These changes are easily depicted by taking the difference of the individual clustered outputs.

All the operations explained in this section are performed individually for each type of texture orientation. For example, if an image has two major regions of different textural orientations, the image will be scanned successively to identify the regions that have differing textural orientations, one after another. If there are no changes over the temporal scale, the clustering algorithm should ideally give similar boundaries for each cluster region. On the other hand, changes in dune orientations will correspond to changes in the boundaries of the obtained cluster regions.

RESULTS AND DISCUSSION

The datasets used for the experiments are summarized in Table 1. Figure 2 and Figure 3 are the images corresponding to the Gobi Desert of central Asia in the China-Mongolia region. Figure 4 and Figure 5 indicate the corresponding clustered outputs obtained using the clustering algorithm mentioned in Section 2. The orange coloured region in Figure 4 and Figure 5 encompasses the desert regions with horizontal orientation of sand dunes, while the blue region indicates the vertically oriented sand dunes. Figure 6 indicates the difference output, obtained using the change detection algorithm in Section 2.

Figure 7 and Figure 8 show the superimposition of corresponding input images and clustered outputs, providing a visualization of the matching between image regions and cluster regions. For the 2nd dataset (Figure 9 and Figure 10), imagery correspond to the Thar Desert region in the north western region of India is used. Figure 11 and Figure 12 show the corresponding clustered outputs respectively. The orange coloured region indicates the desert regions with diagonally oriented sand dunes, while the blue region indicates the sand dunes with horizontal orientation. Figure 13 indicates the difference output. The next dataset (Figure 14 and Figure 15), compromises of another region within the Thar Desert. Figure 16 and Figure 17 show the corresponding clustered outputs for the 3rd dataset. The orange coloured region in Figure 16 and Figure 17 indicates the desert regions with horizontally oriented sand dunes, while the blue region indicates the sand dunes with no particular orientation.

Table 1: Description of datasets

Figure Number	Region	Date	Sensor and Resolution
Figure 1	Gobi Desert	December 1972	Landsat, 30 m
Figure 2	Gobi Desert	January 2011	Landsat, 30 m
Figure 8	Thar Desert Region 1	March 1999	LISS-III, 23.5 m
Figure 9	Thar Desert Region 1	March 2009	LISS-III, 23.5 m
Figure 13	Thar Desert Region 2	March 1999	LISS-III, 23.5 m
Figure 14	Thar Desert Region 2	March 2009	LISS-III, 23.5 m
Figure 18	Sahara Desert	March 1988	Landsat, 30 m
Figure 19	Sahara Desert	March 2005	Landsat, 30 m

Figure 2: Gobi Desert region 1 at the 1st time period.

Figure 3: Gobi Desert region 1 at the 2nd time period.

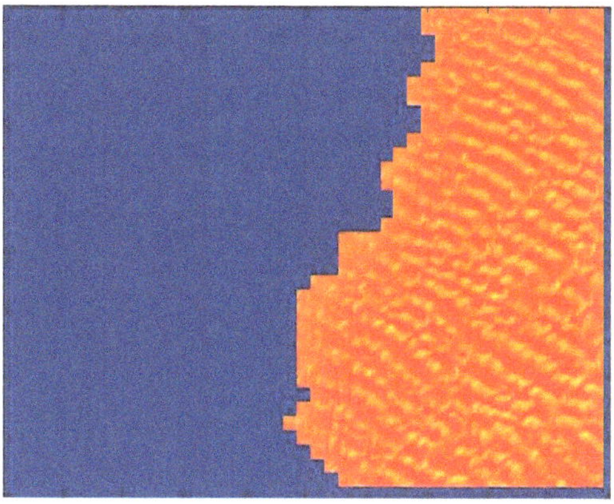

Figure 4: Clustered output of image in Figure 1.

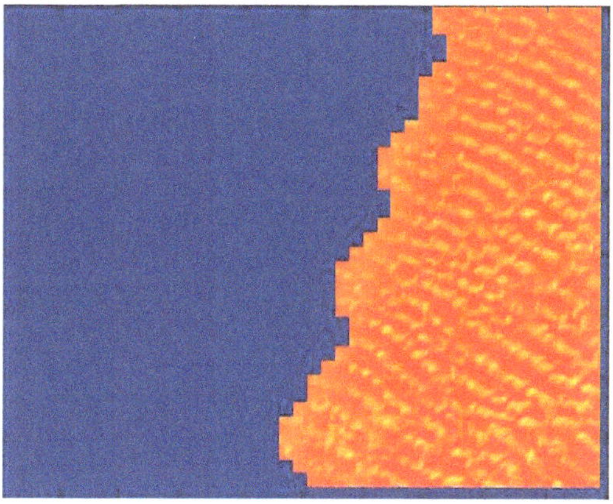

Figure 5: Clustered output of image in Figure 2.

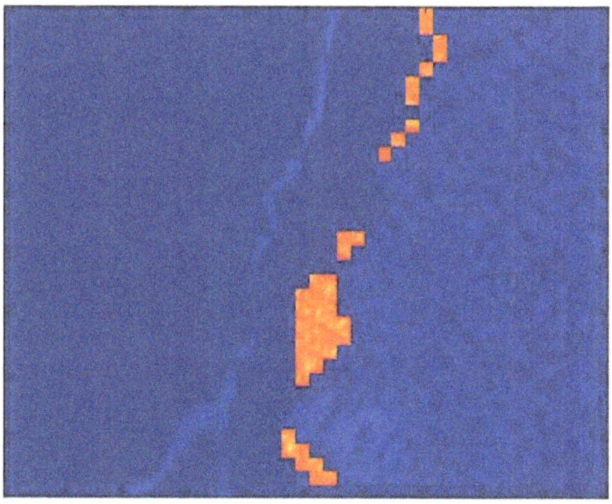

Figure 6: Change detection output: difference output of Figure 4 and Figure 5.

Figure 7: Superimposed clustered output of image in Figure 2.

Figure 8: Superimposed clustered output of image in Figure 3.

Figure 9: Thar Desert region 1 at 1st time period.

Figure 10: Thar Desert region 1 at 2nd time period.

Figure 11: Clustered output of image in Figure 9.

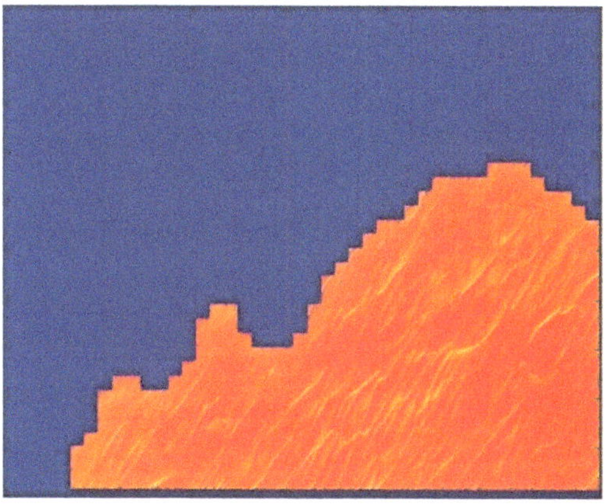

Figure 12: Clustered output of image in Figure 10.

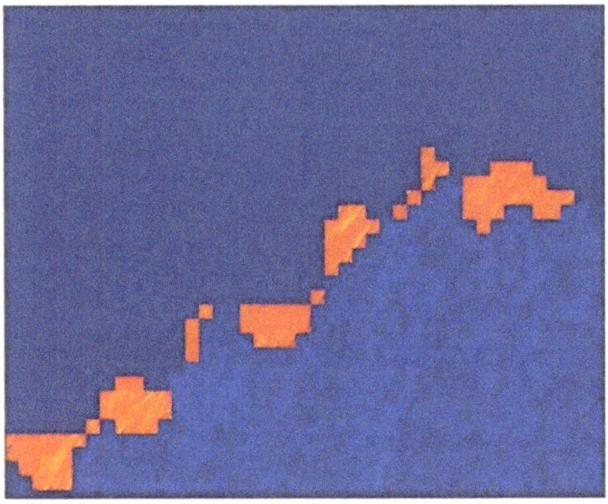

Figure 13: Change detection output: difference output of Figure 11 and Figure 12.

Figure 14. Thar Desert region 2 at 1st time period.

Figure 15: Thar Desert region 2 at 2nd time period.

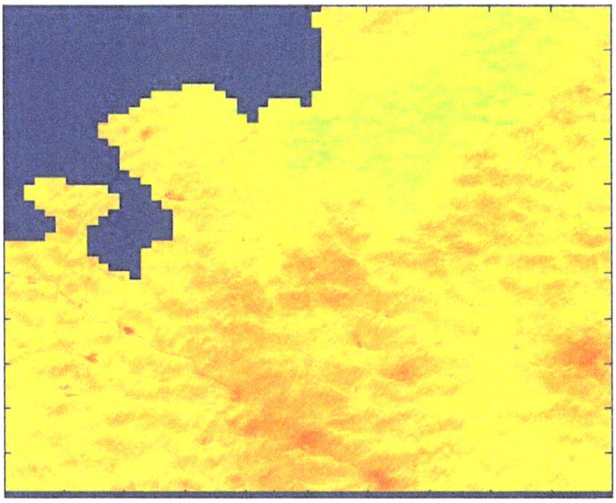

Figure 16: Clustered output of image in Figure 14.

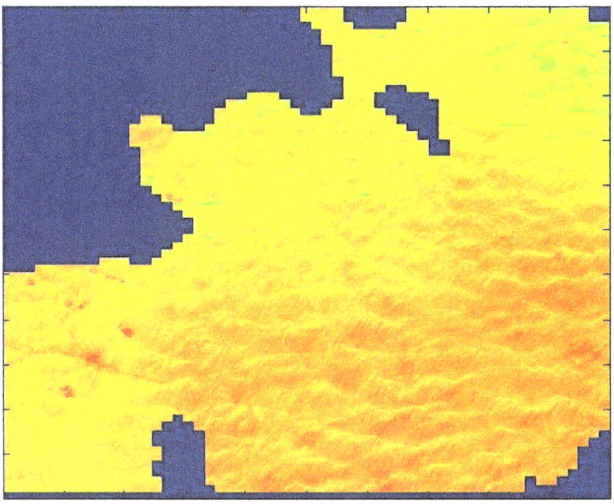

Figure 17: Clustered output of image in Figure 15.

Figure 18 indicates the change detection output for the 3rd dataset. The Sahara Desert region in Yemen is taken as the 4th dataset (Figure 19 and Figure 20) for experimentation. Figure 21 and Figure 22 are the corresponding clustered outputs for this dataset. The orange coloured regions in both these images denote the diagonally oriented sand dunes, while the blue region indicates the absence of dunes with a specific orientation. Figure 23 indicates the change output for this dataset.

The attribute allocation after the Radon transform is performed block wise rather than pixel wise. The block size is dependent on a host of factors, such as the image resolution, size of the texture patterns on the dunes and distance between identically oriented dunes. Larger block sizes have been used to cluster the imagery of the 1st dataset, while smaller block sizes were used to cluster imagery of 2nd dataset.

Optimal thresholds for correct clustering of dunes are decided by observing the texture patterns and comparing them with the generated output regions. It has been ensured that exactly the same thresholds were used for clustering images corresponding to each image set.

The change outputs obtained using methodology adopted to all four datasets indicate that the interiors of the clustered regions are not undergoing any changes. All the minor changes are observed at the boundaries of these clustered regions.

CONCLUSIONS

In this paper, we have proposed a post-clustering methodology to monitor sand dunes, based on changes occurring within cluster boundaries. The knowledge of categorical dune regions and changes occurring over a period of time will aid the administrative setup in taking precautionary measures and thus protect the surrounding ecosystem. The separation of different textured regions was achieved with the use of selective statistical radon features. Change detection of desert sand dunes was performed on desert data sets belonging to three different dune regions. The proposed change detection algorithm worked effectively for the interior parts of imagery, however faltered slightly at the borders of these clusters. This is primarily due to the block based processing technique which enhances the subtle variations at the edges. This block effect is particularly important in the generalization and smoothing of clustered regions.

It has been observed that essentially the dune regions in all the cases remained constant in terms of their orientation structure for three decades. There are only minor changes occurring at the interface of dune patterns. This indicates the stability of wind direction and other factors affecting dune orientations. The steadiness of dune regions with similar textural boundaries over the time periods considered may not necessarily imply the dormancy of sand, however it does point to the similar patterns of wind directions within this period. Hence it can be inferred that over the years, although the actual sand particles may be migrating, there is a definite pattern in their direction of migration.

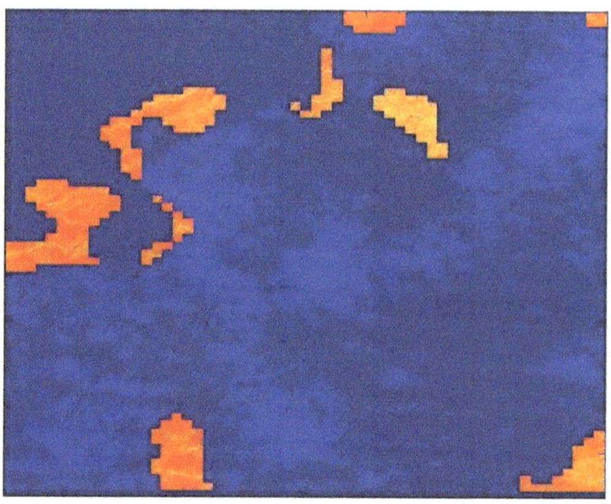

Figure 18: Change detection output: difference output of Figure 16 and Figure 17.

Figure 19: Sahara Desert region at 1st time period.

Figure 20: Sahara Desert region at 2nd time period.

Figure 21: Clustered output of image in Figure 19.

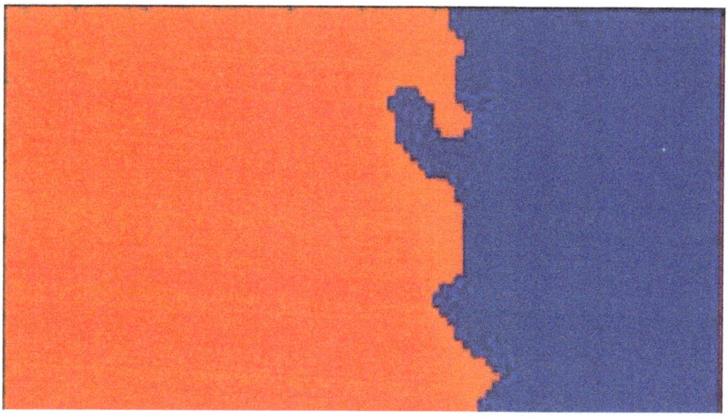

Figure 22: Clustered output of image in Figure 20.

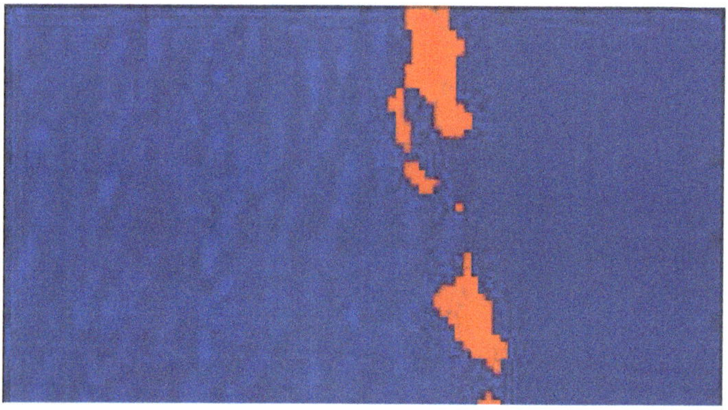

Figure 23: Change detection output: difference output of Figure 21 and Figure 22.

While these findings have implications for understanding of dune geomorphology and the changes occurring in dune directions, they also draw the attention to a wider study of dune orientations and their feedback to associated phenomena both at regional and global scales.

Furthermore, the fact that the proposed block based approach does not require any additional information apart from that available through satellite imagery suggests that it can be used for a host of applications. It must be verified in order to confirm that the steadiness of dune orientations holds well from a historical perspective. Non-availability of open domain remote sensing data prior to 1972 limits the scope of using this methodology for dune change detection.

REFERENCES

1. Belaid, M.A. (2010) Remote Sensing of Desert Cities in Developing Countries. In: Tarek, R. and Carsten, J., Eds., *Remote Sensing of Urban and Suburban Areas*, Vol. 10, Springer Publisher, New York, 245-265.

2. Lancaster, N. (1988) Development of Linear Dunes in the Southwestern Kalahari, Southern Africa. *Journal of Arid Environments*, **14**, 233-244.

3. Paisley, E.C.I., Lancaster, N., Gaddis, L.R. and Greeley, R. (1991) Discrimination of Active and Inactive Sand from Remote Sensing: Kelso Dunes, Mojave Desert, California. *Remote Sensing of Environment*, **37**, 153-166. http://dx.doi.org/10.1016/0034-4257(91)90078-K

4. Yang, X. (2011) Visualization Urban Spatial Growth of Desert Cities from Satellite Imagery: A Preliminary Study. *Proceedings of the 25th International Cartographic Conference*, Paris, 3-8 July 2011, CO-439.

5. Karnieli, A., Gilad, U., Ponzet, M., Svoray, T., Mirzadinov, R. and Fedorina, O. (2008) Assessing Land-Cover Change and Degradation in the Central Asian Deserts Using Satellite Image Processing and Geostatistical Methods. *Journal of Arid Environments*, **72**, 2093-2105. http://dx.doi.org/10.1016/j.jaridenv.2008.07.009

6. Aldossary, A. (2012) Analysis of Urban Change Detection Techniques in Desert Cities Using Remote Sensing. Master's Thesis, California State University, Northridge.

7. Radon, J. (1917) On the Determination of Functions from Their Integrals along Certain Manifolds. *Mathematische- Physikalische*, **69**, 262-267.

8. SurenderVarma G., Banerjee, B. and Buddhiraju, K.M. (2012) Representation of Desert Sand Dunes by Surface Orientation Using Radon Transform. *Proceedings of IGARSS*, Munich, 22-27 July 2012, 2733-2736.

9. Jafari-Khouzani, K. and Soltanian-Zadeh, H. (2005) Radon Transform Orientation Estimation for Rotation Invariant Texture Analysis. *IEEE Transactions on Pattern Analysis & Machine Intelligence*, **27**, 1004-1008. http://dx.doi.org/10.1109/TPAMI.2005.126

10. Haralick, R.M., Shanmugam, K. and Dinstein, I. (1973) Textural Features for Image Classification. *IEEE Transactions on Systems, Man and Cybernetics*, **6**, 610-621. http://dx.doi.org/10.1109/TSMC.1973.4309314

4
CHAPTER

ANALYSIS OF MOUNTAIN RANGELAND CHANGES, USING REMOTELY-SENSED DATA

Ali A. Torahi and Elham Manuchehri

Keywords: Rangeland, Changes, Remote Sensing, Dehdez, Iran

INTRODUCTION

On an area basis, rangelands contribute significantly to the world's land surface and an important share is devoted to grazing. Management of these areas drastically alters their natural characteristics [1]. With increasing livestock, pressure on natural and semi-natural rangelands also increases. However, not all range types have the same production potential. Destruction of rangeland is mainly related to overpopulation. To enhance food production, grasslands possessing fertile soils have been ploughed and converted to agricultural lands [2]. Three processes, desertification, woody encroachment, and deforestation, are inherent to global grazing. These processes have widespread and differential effects on the structure, biochemistry, hydrology and biosphere-atmosphere exchange of grazed ecosystems, as well as representing a major component of global environmental change [3]. Managing rangelands represents a major shift in thinking and practice, many believe that this is the required stimuli to develop a sustainable rangelands management approach [4]. There is an increasing need for improved management of rangelands in developing countries due to the alarming depletion caused by population pressure, agricultural expansion, and the misuse of rangelands [5]. The implementation of any rangeland management strategy, whether it is for the assessment and allocation of sustainable use or for rehabilitation of rangelands and denudate lands, requires a quantification of natural resources, their location and their present condition [6]. Knowledge of the rangelands and their geographical environment is essential for

Citation: Torahi A.A. and Manuchehri E. (2016) Analysis of Mountain Rangeland Changes, Using Remotely-Sensed Data, Journal of Geoscience and Environment Protection, 4, 100-106. http://dx.doi.org/10.4236/gep.2016.41012

proper planning of sustainable rangeland management. To meet these requirements, precise and up-to-date information regarding the status of the rangelands and the potential for rehabilitation is important to upgrade and to design suitable management planning for future improvement of the rangelands [7]. Several studies have reported successful mapping of rangelands in arid and semi-arid environments, as well as in temperate areas, based on remotely sensed data [8]. In tropical and subtropical areas, attempts to classify land-cover have previously been carried out [9]. In this study area, raring cattle is a very important economic activity and continuous grazing all year round is possible at almost all sites [10]. Hence, precise land-cover information as well as a quantification of the study area is required. The present study was undertaken in the Dehdez area to assess the trends of rangeland dynamics in the study area during the period of 1990-2006.

STUDY AREA

The study was carried out in Dehdez area, located in the southern part of Iran, specifically in the south-eastern Khuzestan province, between 50°12'14" to 50°33'25"E and 31°35'05" to 31°58'12"N (Figure 1). The area consists of hills and mountains and some gentle slopes. The altitude ranges from 580 to 3000m asl and covers an area of 512 km^2. Three forest types dominate the area: *Amygdalus scoparia*, *Quercus persica* and *Pistacia attlantica*. *Quercus* sp. (*Quercus persica*) is a pioneer species and is the dominant forest type in this area [10]. The area is covered by forest, shrub land, agriculture, settlements, villages, riverbeds and water bodies. The main soil types in the area consist of Mollisols, Entisols, Inseptisol and Alfisols [11]. Different lithology in parent material on mountains and hills is observed, creating a landscape of varying slopes. The steep slopes are highly affected by erosion, with shallow soil that have a high drainage and coarse texture.

METHODOLOGY

Image Pre-Processing

Two clear, cloud-free Landsat images and one ASTER images were selected to classify the study area on June 17, 1990; May 18, 1998 and June 21, 2006. respectively. The Dehdez area is entirely contained within the Landsat and ASTER path of 164 and row 38. All images were rectified to UTM zone 39, WGS84 using at least 25 well distributed ground control points and nearest neighbour resampling.

Figure 1: Location of Dehdez in Khuzestan province, Iran.

The root mean square errors were less than 0.25 pixels (7.5 m) for each of the three images. Image processing was performed using ENVI 4.5. Land-use/cover mapping is achieved through the interpretation of satellite imagery

Image Classification

Training

The classification scheme was based on the land cover and land use classification system developed by [12] for the interpretation of remote sensing data at various scales and resolutions. Based on the Anderson land-use/cover classification system, the land-use and land-covers are classified as forest land, rangeland, water bodies, agricultural land and residential land. An unsupervised image classification method was carried out prior to the field visit, in order to determine strata for ground truthing. Data was collected in the field in order to train and validate the land-use/cover interpretation of the satellite image from 2006, and for the qualitative description of the characteristics of each land-use/cover class. Additional ancillary and historical data was required for classification of the 1990 and 1998 images [13]. In order to create a sample testing set, a set of testing points was randomly selected [14]. However, reaching all the random points in practice is infeasible, as the study area is very complex with steep slopes, and hence difficult to access, particularly in areas that are very far from the road/path. Therefore, a modification was made in the field, whereby 95 random points were used and the cover classes that were mapped in the vicinity of these points were checked. For the classification of images of 1990 and 1998, simple random sampling is applied [14].

Allocation

A supervised classification technique with a maximum likelihood algorithm was applied. The classification was based on 48 training samples for the image of 2006, and 42 samples for the images of 1990 and 1998. The land-use/cover maps of 1990, 1998 and 2006 were created using the supervised image classification technique based on the maximum likelihood classifier (MLC) and 132 training samples [15]. Finally, a 3x3 majority filter was applied to each classification, to recode isolated pixels classified differently compared to the majority of the pixels within the window of the class.

Testing

An independent sample of an average of 95 polygons, with approximately 100 pixels for each selected polygon, was randomly selected from each classification to assess classification accuracies [16]. Error matrices, given as cross-tabulations of the mapped class vs. the reference class, were used to assess the classification accuracy [17]. Overall accuracy, user's and producer's accuracies, and the Kappa statistic were then derived from the error matrices.

RESULTS AND DISCUSSION

The land-use/cover pattern in the area as a whole showed that the rangeland covers about 30.8%, 36.7% and 45% of the total geographical area of the Dehdez area in 1990, 1998 and 2006, respectively. According to the confusion matrix, the overall accuracies of land-use/cover classification for 1990, 1998 and 2006 were 89.37%, 75.24% and 71.14%, respectively. The Kappa statistic was made for verifying the accuracy of land-use/cover maps. The Kappa values obtained were of 78.71%, 55.61% and 51.41% of accuracy for the years 1990, 1998 and 2006, respectively. During the 16 years span period (1990-2006), approximately 1738.4 ha, 383.7 ha, 32.8 ha and 890.1 ha of rangelands were converted to forest, agriculture, water and settlement respectively. However, due to the of conversion of other land covers to rangeland, the total rangeland increased by 14.2% during 1990-2006 (Figure 2). The rate of rangeland increment was approximately 1.77% per year (Table 1). There are two types of rangelands in Dehdez area, rich rangeland and poor rangeland. The monitoring of the spatial distribution of the rangelands revealed that the northern and some eastern parts of the area was dominated by the rich rangelands (Figure 3). Total rich rangelands in the area accounted for 38.5%, 44% and 42.2% in 1990, 1998 and 2006, respectively (Table 2). Due to the steep slopes and the inaccessibility of these areas, the rangelands have s more species diversity compared to the other land types. The poor rangelands were dominated in southern and eastern parts of the area (Figure 3). The total poor rangeland in the area accounted for 61.5%, 56% and 57.8% in 1990, 1998 and 2006, respectively (Table 2). These rangelands are used as pasture lands, with the majority degraded by livestock overgrazing. Grazing occurred wherever poor rangelands were found to be accessible. In general, poor rangelands close to settlements experience intensive grazing. High degradation is visible in the poor rangelands within the study area, and species that are able to withstand grazing have crowded out original species in many areas.

Table 1: Rangeland changes from the land-cover change detection analysis in the Dehdez region

Change type		1990-1998		1998-2006		1990-2006	
		Area		Area		Area	
From	To	(ha)	(%)	(ha)	(%)	(ha)	(%)
Rangeland	Forest	3111.6	19.7	4390.8	23.3	1738.4	11.0
Rangeland	Agriculture	636.8	4.0	435.7	2.3	383.7	2.4
Rangeland	Water	481.5	3.0	42.6	0.4	32.9	0.2
Rangeland	Settlement	983.1	6.2	1512.0	8.0	890.1	5.6

Table 2: Area under different rangeland types in the Dehdez area

Year / Rangeland type	1990		1998		2006	
	Area		Area		Area	
	(ha)	(%)	(ha)	(%)	(ha)	(%)
Rich rangeland	6094	38.5	8294	44.0	9762	42.2
Poor rangeland	9720	61.5	10545	56.0	13350	57.8
Total	15814	100	18839	100	23112	100

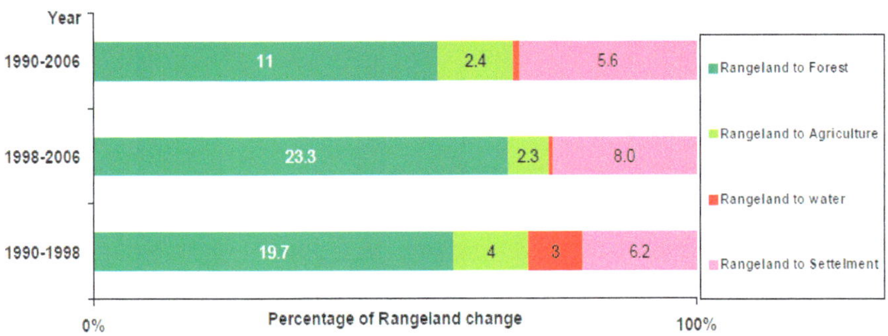

Figure 2: Rangeland changes in different directions as a percentage of the total rangeland area for the years of 1990-1998, 1998-2006 and 1990-2006.

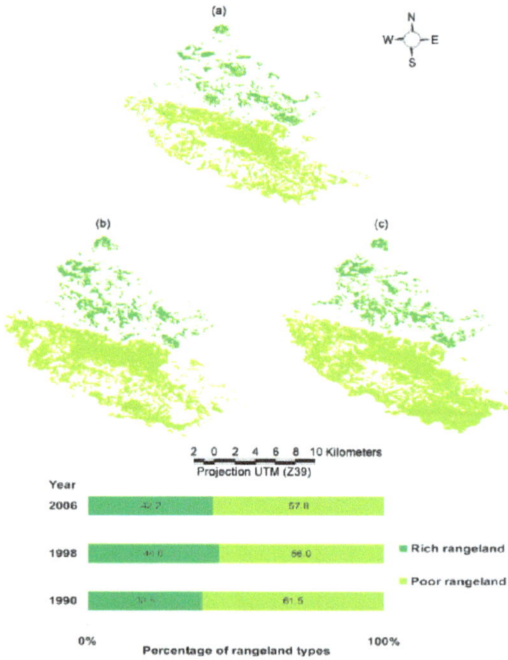

Figure 3: Rangeland in the Dehdez area: (a) 1990; (b) 1998 and (c) 2006 and a comparison of respective extents of rangeland types by percentage in the area.

CONCLUSION

Satellite remote sensing enables the generation of a detailed rangeland map, and the separation of grazing intensity levels in rangelands can be generated with the relatively little effort in areas that are difficult to access [18]. It also enables a more realistic assessment of livestock grazing pressure in the area, and sets a framework for future management planning. Approximately 36.6% of Dehdez's area of rangeland shows mismanagement features and almost 26% is classified as intensively or overgrazed. Based on land-cover dynamic mapping, an increase in the area of rangelands during the period of study was observed, however because of the high pressure of human activity, the quality of these rangelands is seriously affected. Many changes have occurred in the rangelands due to, for example human activities, domestic animals and ecological conditions. Human activity is the primary factor affecting the rangelands in the study area. Animal husbandry leads a high consumption of grass via the frequent pasturing by domestic animals. Agricultural activities, such as deep plaguing, decrease in the efficiency of the soil in the area. In addition, the rangelands are exploited for medicinal plants, as well as for fuel and high quality wood. These activities further damage rangelands, with the quality of vegetation cover, even as food and fodder, is reduced. The damaging union of these factors is adversely affecting the ecological cycle of this land type. Natural rangelands are consequently changing to agricultural land types and moving away from natural vegetation cover. The quality of vegetation cover is decreasing due to unbalanced conditions, for example the unequal ratio of vegetation cover and animals. The lack of plant cover due to overgrazing causes soil erosion by water and wind. In conclusion, efforts should be made to prevent degradation and deterioration of rangelands, irrespective of scale and geographic location.

REFERENCES

1. Asner, G.P., Elmore, A.G., Olander, L.P., Martin, R.E. and Harris, A.T. (2004) Grazing Systems, Ecosystem Responses, and Global Change. *Annual Review of Environment and Resources*, 29, 261-299. http://dx.doi.org/10.1146/annurev.energy.29.062403.102142
2. Cook, F.M. (1999) The Challenge of Sustainable Forests. Forest Resources Policy in Malaysia, 1970-1995. University of Hawaii Press, Honolulu.
3. Laliberte, A.S., Rango, A., Havstad, K.M., Paris, J.F., Beck, R.F. and McNeely, R. (2004) Object-Oriented Image Analysis for Mapping Shrub Encroachment from 1937 to 2003 in Southern New Mexico. *Remote Sensing of Environment*, 93, 198-210. http://dx.doi.org/10.1016/j.rse.2004.07.011
4. Evans, J. and Geerken, R. (2006) Classifying Rangeland Vegetation Type and Coverage Using a Fourier Component Based Similarity Measure. *Remote Sensing of Environment*, 105, 1-8. http://dx.doi.org/10.1016/j.rse.2006.05.017
5. Sedano, F., Gong, P. and Ferrao, M. (2005) Land Cover Assessment with MODIS Imagery in Southern African Miombo Ecosystems. *Remote Sensing of Environment*, 98, 429-441. http://dx.doi.org/10.1016/j.rse.2005.08.009
6. David, S.W. and John, T.F. (1996) Remote Sensing Imagery for Natural Resources Monitoring—A Guide for First Time Users. Columbia University Press, New York.
7. Richards, J.A. (1993) Remote Sensing Digital Image Analysis. An Introduction, 2nd Edition, Springer-Velarg, Berlin Heidelberg. http://dx.doi.org/10.1007/978-3-642-88087-2
8. Serneels, S. (2001) Divers and Impact of Land-Use/Land-Cover Change in the Serengeti-Mara Ecosystem: A Spatial Modeling Approach Based on Remote Sensing Data. Universite Catholique de Louvain, Belgium, Louvain.
9. Lillesand, T.M. and Kiefer, R.W. (2000) Remote Sensing and Image Interpretation. 4th Edition, Wiley & Sons, New York, 724.
10. Fattahi, M. (1995) Investigation of Zagros Natural Resources and the Important Factors of Demolition. Institute of Forests and Range Lands Researches, 57.
11. Piraste (2004) Use of Remote Sensing and GIS to Interprate Structural and Tectonic of Zagros Belt-Southweast Iran. Thesise of PhD, Aligarh Muslim University, 197.
12. Anderson, A. (1976) Land Use and Land Cover Classification System. Geological Survey Professional Paper 964, Washington DC.

13. Serneels, S., Said, M.Y. and Lambin, E.F. (2000) Land-Cover Changes around Major East Afrian Wildlife Reserves: The Mara Ecosystem. *International Journal of Remote Sensing*, 22, 3397-3420. http://dx.doi.org/10.1080/01431160152609236
14. Jensen, J.R. (1996) Introductory Digital Image Processing. Trentice Hall, New Jersay.
15. Lunettaa, R.S., Ediriwickremba, J., Johnsonb, D.M., Lyonc, J.G. and McKerrowb, A. (2002) Impacts of Vegetation Dynamics on the Identification of Land-Cover Change in a Biologically Complex Community in North Carolina, USA. *Remote Sensing of Environment*, 82, 258-270. http://dx.doi.org/10.1016/S0034-4257(02)00042-1
16. De Gier, A., Westinga, E., Beerens, S., van Laake, P. and Savenije, H. (1999) User Requirements Study for Remote Sensing-Base Spatial Information for the Sustainable Management of Forests—Final Report: ITC; FAO; National Reference Centre for Nature Management [IKC Natuurbeheer].
17. Congalton, R.G. and Green, K. (1999) Assessing the Accuracy of Remotely Sensed Data: Principles and Practices. Florida' Lewis Publishers, Boca Rotan, 43-64.
18. Petit, C., Scudder, T. and Lambin, E.F. (2001) Quantifying Processes of Land-Cover Change by Remote Sensing: Resettlement and Rapid Land-Cover Changes in South Eastern Zambia. *International Journal of Remote Sensing*, 22, 3435-3456. http://dx.doi.org/10.1080/01431160010006881

5
CHAPTER

CONTINENTAL COLLISION STRUCTURES AND POST-OROGENIC GEOLOGICAL HISTORY OF THE KANGERLUSSUAQ AREA IN THE SOUTHERN PART OF THE NAGSSUGTOQIDIAN OROGEN, CENTRAL WEST GREENLAND

Jon Engström and Knud Erik S. Klint

Keywords: Continental collision structures; Kangerlussuaq–Russell thrust fault; tectonic structural model; Nagssugtoqidian orogeny; West Greenland

INTRODUCTION

The Nagssugtoqidian orogeny of West Greenland occurred predominantly between 1.92 and 1.75 Ga [1], as a consequence of the convergence and collision of the North Atlantic Craton and a poorly defined craton to the north [2]. The Palaeoproterozoic reworking during the orogeny extends from Kangerlussuaq to southern Disko Bugt [3–5]. Several studies have been performed in the central and northern areas of the Nagssugtoqidian orogen (e.g., [1,6]), while the southern end of the orogen at the northern margin of the North Atlantic Craton, in the proximity of Kangerlussuaq, has largely remained unexplored. However, since 2008, the international Greenland Analogue Project (GAP) has been conducting intensive research in the area around Kangerlussuaq [7]. The main goal of the GAP has been to investigate the deep groundwater in glaciated terrains. In order to study and model the hydrogeology, a detailed geological investigation and a conceptual geological 3D model of the fault/fracture zones is required. The GAP drilled three cored boreholes in the vicinity of the ice margin, which provided data for the 3D geological interpretation. The geological survey revealed a number of structures formed

Citation: Jon Engström and Knud Erik S. Klint, Continental Collision Structures and Post-Orogenic Geological History of the Kangerlussuaq Area in the Southern Part of the Nagssugtoqidian Orogen, Central West Greenland, doi:10.3390/geosciences4040316

in a plate tectonic collision zone, and the main goal of this paper is to add complementary information to aid in understanding the formation of the Nagssugtoqidian orogen in central West Greenland.

BACKGROUND: THE DEVELOPMENT OF THE NAGSSUGTOQIDIAN OROGEN AND THE STUDY AREA

The Nagssugtoqidian orogen is interpreted as a collisional orogenic setting, where juvenile Palaeoproterozoic igneous rocks intruded Archaean units and where Palaeoproterozoic sedimentary volcanic rocks were deposited. The orogen is now seen as an approximately 250-km-wide, south- to south-eastward-trending belt that includes diverse Archaean and Palaeoproterozoic rocks affected by polyphase deformation and high-grade metamorphism [8]. Along the northern margin of the North Atlantic craton, within the southern Nagssugtoqidian orogen, Archaean dioritic, granodioritic, tonalitic and granitic orthogneiss of the North Atlantic craton predominate, albeit in a reworked state [2,9–12]. The orthogneisses have been dated to 2.87–2.81 Ga, with the plutonic protoliths deformed and metamorphosed between 2.81 and 2.72 Ga, immediately after their emplacement [11,12]. The orthogneisses are cut by several sets of mafic dykes, the most voluminous of which is the E–W- to NE–SW-trending, 2.05–2.04 Ga Kangâmiut dyke swarm [13–21]. The Kangâmiut dyke swarm, which extends from the Archaean foreland into the centre of the orogen, has become reworked across a sharp transition at the southern Nagssugtoqidian front. Its northern extent is abruptly terminated by the Ikertoôq thrust zone (Figure 1), which is a south-vergent ductile shear zone [2]. Dykes from the Kangâmiut swarm occur in the Archaean gneiss complex at approximately 150 km south of the thrust zones, traditionally taken as marking the southern Nagssugtoqidian front in SW Greenland, and continue as deformed bodies in the reworked Archaean gneisses for approximately 50–100 km to the north of the front [17].

The granulite-facies thermal peak of metamorphism in the core of the orogen was dated at *ca.* 1.86–1.84 Ga [11,20,22,23]. This age interval coincides with the timing of thrusting and is interpreted as documenting the main phase of collision across the Nagssugtoqidian orogen. The Nagssugtoqidian orogeny is interpreted as being associated with the closure of two south-dipping Palaeoproterozoic sutures [6,24,25]: (1) a northern suture, the Disko Bugt suture, which separates the lower plate Rae craton from the middle Aasiaat domain; and (2) a southern suture rooting in the Nordre Isortoq steep belt that separates the southerly-located North Atlantic Craton from the Aasiaat domain [26]. This model was further developed by Garde and Hollis [1], who concluded there are two south- to SSE-dipping subduction zones in the central and northern Nagssugtoqidian orogen, which were separated by the previously unidentified Aasiaat domain or microcontinent.

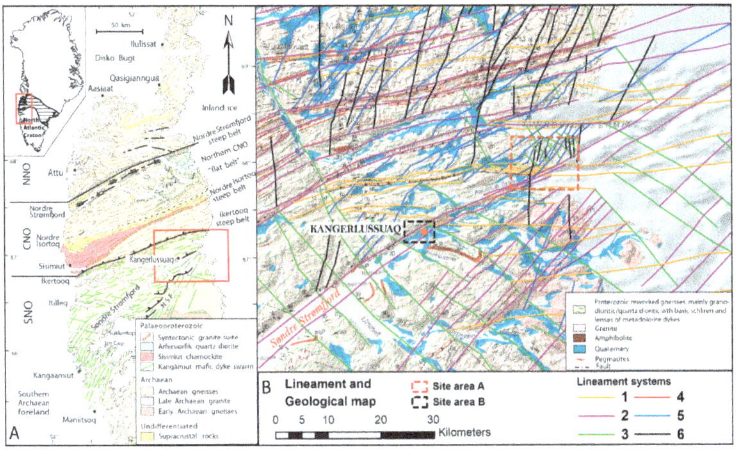

Figure 1: (**A**) Geological map of central West Greenland and the Nagssugtoqidian orogen, indicating the location of Kangerlussuaq. Map modified from van Gool *et al.* [2]. (**B**) A lineament and geological map of the study area, in front of the Greenland ice sheet. Map modified from Garde and Marker [27].

The Study Area

The study area is located at the southern end of the orogen in the vicinity of the North Atlantic Craton

(Figure 1). It covers a domain of 60 km × 80 km, with a special focus on an area at the front of the Greenland ice sheet (Figure 1A). Most of the mapping, including the drilling of three deep boreholes, was conducted in area A (Figure 1B) during field campaigns in 2008–2009, while area B was surveyed during mapping campaigns in 2010–2011. A geological 3D model close to the boreholes was constructed for groundwater modelling purposes. The mapping at other locations primarily focused on the nature of the lineaments and their crosscutting relations, in order to identify lineaments with potential for hydraulically conductive fractures/faults. A preliminary geological event succession based on lineament interpretation of the area was presented by Klint *et al.* [28] in 2013, complementary data was collected in the field, and this paper presents an updated geological model of the Kangerlussuaq area.

The study area, extending from Kangerlussuaq Airport to the ice sheet, is dominated by primarily felsic banded gneisses and mafic garnet-rich orthogneisses, intruded in places by pronounced 1–25-m-wide Kangâmiut dykes, especially in the southwestern part of the study area. Detailed geological mapping focused on selected sites from Kangerlussuaq Airport to the margin of the ice sheet (Figures 1B and 2). The Nagssugtoqidian structures reflect the general ductile nature of the regional deformation and include a penetrative gneissic fabric, macro-scale folds, and evidence of shearing in pronounced shear zones. Brittle structures, such as faults and fractures, are abundant and were probably formed in a younger, shallow, colder and hence more rigid environment.

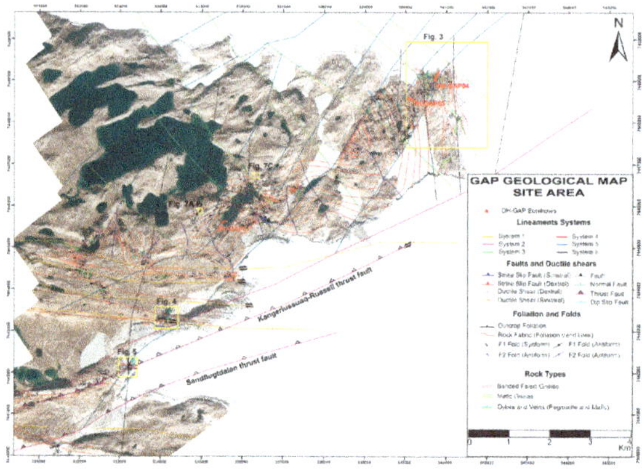

Figure 2: Geological map illustrating the distribution of the lineaments in the ice-margin valley area, as well as the main general structures and rock fabric. Seven key locations were investigated in 2011 and three more (Figures 3–5) in 2013. The key locations described below are highlighted along with the different lineament systems defined by Klint *et al.* [28].

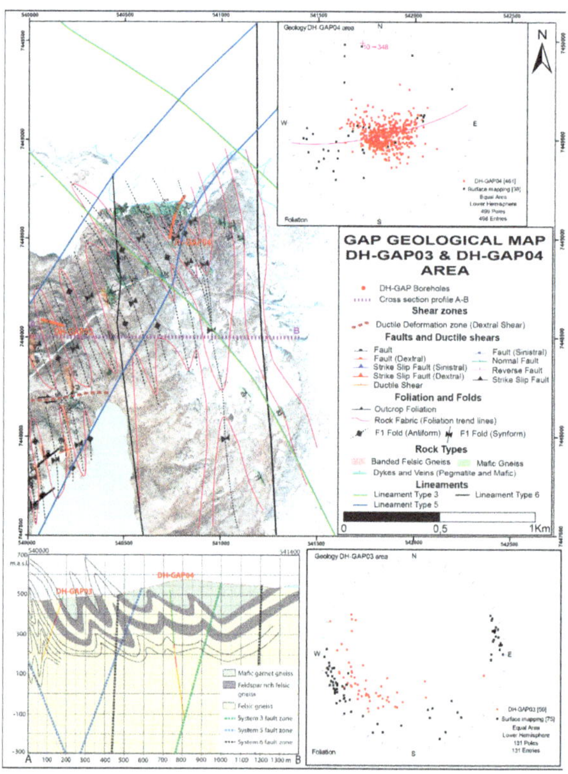

Figure 3: Geological map illustrating the geology around the two boreholes, DH-GAP03 and DH-GAP04. The cross-section illustrates how the folding pattern changes from a larger open fold in the east towards a tighter isoclinal folding in the west. The same can be observed in the stereographic plots.

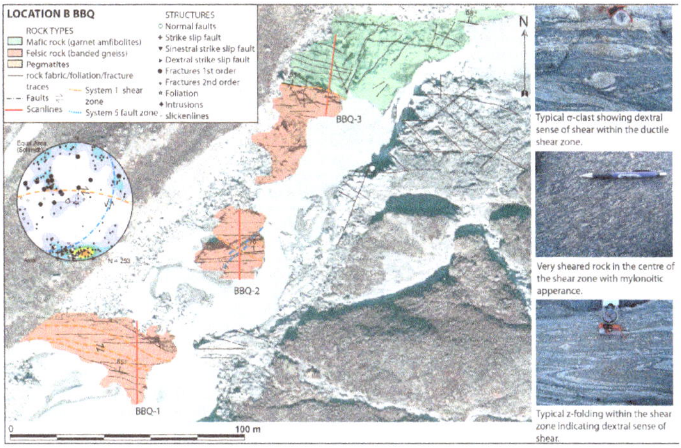

Figure 4: Location B. Overview of the BBQ location. Detailed measurements of faults and fractures were conducted along 3 transects, BBQ 1–3. Selected outcrops with structures showing cross-cutting relationships were measured and the event succession of the structures was evaluated.

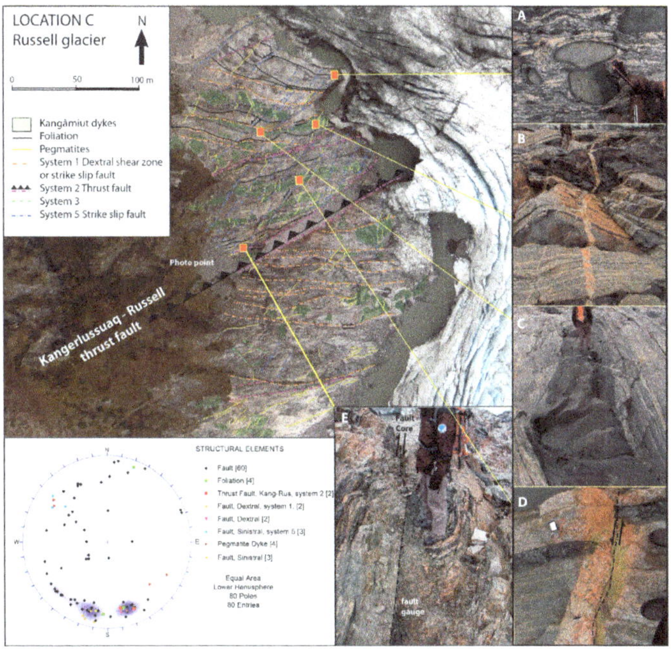

Figure 5: Cross-cutting relations of systems 1, 2 and 5, and deformation style. (**A**) Intensely deformed gneiss with inclusions of mafic rocks of unknown origin. (**B**) Banded felsic and mafic gneiss truncated by pegmatites. (**C**) A folded Kangâmiut dyke. (**D**) System 5 sinistral strike-slip faults with syntectonically formed pegmatites. (**E**) The fault core in the Kangerlussuaq–Russell thrust fault. Multiple thrust planes and crushed zones with fault gauge.

REGIONAL LINEAMENT MAPPING

Most lineaments in crystalline rocks represent structural features such as faults, shear zones, rock fabric, or less competent zones reflecting differences in rheology. Lineament mapping generally comprises of three steps: (1) identification of lineaments using remotely sensed GIS data, compiled from aerial photographs, together with topographical, geological and geophysical data; (2) selection of lineaments and verification of their nature and individual cross-cutting relations in the field (descriptive and kinematic analysis); and finally (3) the development of a local event succession model by determining overprinting relationships [28].

According to Klint *et al.* [28], five lineaments primarily representing fault zones were outlined from the GIS analysis (systems 1, 2, 3, 5, 6). The subsequent fieldwork in 2011 confirmed these lineaments, and also identified a sixth sub-horizontal fault system (system 4). The curvilinear nature of the foliation traces highlighting the superposition of two stages of folding (F1 and F2) was also investigated (Figure 2).

During the fieldwork in 2013, cross-cutting relations between the different lineament systems were established and the major lineament extending from Kangerlussuaq to the Russell Glacier, hereafter denoted as the Kangerlussuaq–Russell thrust fault, was mapped in more detail.

The lineaments were correlated by Klint *et al.* [28] as follows:

System 1 lineaments (orange) are orientated E–W, with a steep dip towards N (Figure 2). Locally, these lineaments dominate the southern part of the field area and consist of ductile E–W-trends, locally 10–100-m-wide, foliation-parallel, steeply N-dipping dextral shear zones (Figures 2 and 4). During the semi-ductile shearing of these steep zones, it is recognised that the Kangâmiut mafic dykes were sheared and deformed, thus indicating that these shear zones are younger than the 2.05–2.04 Ga Kangâmiut dyke swarm [21]. Immediately north of the study area, the prominent Ikertoôq thrust zone has been described by van Gool *et al.* [2] and Grocott [29] as a two-phased deformation zone or shear belt, with the superposition of two simple shear strains with varying orientations. The earlier simple shear strain initiated the shear belt and was characterized by horizontal ENE–WSW movements in a vertical shear plane. The second deformation was produced by over-thrusting in a SSE direction [29]. The system 1 lineaments in this area are interpreted to be similar to the shear belts further north, showing the same deformation style as the first phase of deformation in the thrust zone further north.

System 2 lineaments (lilac) are the most pronounced lineaments in the study area. They represent regional SW-trending shear zones, including two zones that are parallel to each other. The first of these is presently occupied by Søndre Strømfjord and stretches along the northern margin of Sandflugtdalen (translated as the Sand drift valley), hereafter named the Sandflugtdalen thrust fault. The second is situated to the north, stretching from Kangerlussuaq to the Russell Glacier, hereafter named the Kangerlussuaq–Russell thrust fault (Figure 2). Further north (Figure 1A), the Nordre Isortoq and Nordre Strømfjord steep belts [2] outline similarly oriented lineaments. In the Kangerlussuaq–Russell thrust fault, this lineament is dominated by dextral strike-slip movement overprinting elements of thrust faulting dipping towards the north. Large boudins of deformed Kangâmiut mafic dykes [21] are cross-cut by this fault zone, which is accordingly younger than the intrusion of the Kangâmiut dykes. The second deformation event in the Ikertoôq thrust zone was produced by thrusting towards SSE [29], which is similar to the thrusting occurring in the Kangerlussuaq–Russell thrust fault.

System 3 lineaments (green) are a major system trending NW, outlining pronounced valleys all the way up to Aassiaat and the Disko Bay. Within this area, they form semi-brittle sinistral strike-slip fault zones that displace the system 1 shear zones.

System 4 (red) forms oblique to sub-horizontal fracture systems trending NNE and is not represented by any pronounced lineaments. These structures may also be related to a population of normal faults and younger NNE-trending pegmatites oriented parallel to the normal faults. They overprint system 1 and 3 structures, but no cross-cutting relationship with system 2 structures has been observed.

System 5 (blue) generally consists of semi-brittle sinistral strike-slip faults trending NE and dipping towards NW or SE. This system outlines the pronounced escarpment in the central-north-eastern corner of the area (Figure 2). Locally, zones of thrust faulting striking WSW and dipping north and south are regarded to represent local transpression in a wrench fault system related to the overall system 5 lineaments [30]. This system is often related to pegmatites, which strike parallel to the faults.

System 6 (black) outlines N–S-trending brittle dextral strike-slip faults. This system may be related to system 5 as a conjugate system. The age of this deformation is questionable; however, lineaments with the same orientation near the west coast have been related to young tectonic events during the rifting between Greenland and Canada that took place during the last 100 Ma [30].

FIELD OBSERVATIONS FROM SITE AREAS A AND B

The Kangerlussuaq area may be separated into two sub-regions: the ice margin valley (site area A) and the Kangerlussuaq Airport area (site area B) (Figure 1B).

The mapped and interpreted structures (foliations, folds, hinge lines, faults and shear zones) are shown on a high-resolution aerial image to provide an overview of the geological framework of the study area (Figure 2). Close to the ice sheet margin, in the north-eastern part of the area, the bedrock is characterised by folded mafic garnet gneiss that defines an open, NNW-trending and shallowly plunging (*ca.* 10°) F1 fold structure. The F1 folds gradually turn into tight/isoclinal steeply dipping overturned folds that can be traced south and westward around an open and steeply plunging, kilometre-scale F2 fold structure with a steeply dipping fold axis (*ca.* 70–60°) towards N–NW. The large-scale F2 fold structures are repeated several times, going from east to west in the northern part of the study area. However, the southern part of the study area is dominated by the aforementioned (orange) E–W-trending lineaments, forming 10–100-m-wide foliation-parallel semi-ductile shear zones. The rock type to the west and south is mainly felsic orthogneiss, with varying amounts of garnet. The 10–100-m-wide foliation-parallel shear zones (orange lineaments) are believed to be syntectonic with the formation of the F2 folding. These shear zones may have acted as the décollement plane for the folding. A few key locations (locations A, B, C and D) were mapped in detail to enhance knowledge of the geology and to determine the cross-cutting relationships.

Location A: Geology around Boreholes DH-GAP03 and DH-GAP04; F1–F2 Folding, System 3, 5 and 6 Lineaments

The area close to location A has been subjected to multiple phases of folding, where the F1 folds usually have a tight to isoclinal pattern. This pattern is widely distributed over the northern part of site area A, with the exception of the north-eastern part, where the structures are dominated by an open F1 fold in mafic garnet gneiss (Figure 3). This fold shallowly plunges (*ca.* 10°) towards the NNW. It has been recognised as a structure extending down into the bedrock, as demonstrated by the deepest research borehole (DH-GAP04), showing that the fold extends to a vertical depth of at least 650 m. To the west, this shallowly plunging fold turns into a tight/isoclinal folding pattern with several standing/overturned folds having a wavelength of approximately 50–100 m, stretching over a 0.5-km-wide area around DH-GAP03 (Figure 3). These folds also steeply plunge towards the NNW, and to the south of DH-GAP03 they are intersected by an E–W-trending 50–100-m-wide ductile shear zone with a dextral sense of shear (Figure 3).

Further to the southwest of this area, the F1 folds gradually turn into an open and steeply-plunging, kilometre-scale F2 fold structure with a WNW-trending axial plane (Figure 2). Further west, the F2 folding is more frequent and also becomes tighter, and furthest to the west the folds are tight to isoclinal, overturned and tilted towards E, and with the fold axis dipping steeply towards the N–NW (Figure 2).

Location B, System 1 and 5 Lineaments

Location B is situated in the southern part of the ice-margin valley, which is less folded than the previously described northern part. The area is generally dominated by E–W-trending foliation steeply dipping towards the N. The bedrock is truncated by 10–100-m-wide foliation parallel shear zones (orange system 1 lineaments), which are found at varyingly spaced intervals along a 2-km-wide zone (Figure 2). A number of outcrops outlining well-exposed sections of the bedrock are situated in the riverbed following the northern flank of the Russell Glacier. A total of six outcrops were investigated in this riverbed, and for each of them an event succession was outlined in order to reconstruct the tectonic history of the area [28]. Location B was selected to represent the system 1 and 5 lineaments. It is situated on the riverbed close to the Russell Glacier in the southern part of the geological map (Figure 2).

The dominant rock types constitute a sequence of felsic banded gneisses at least 150 m thick, striking E–W and dipping 86–65° towards the N. More than 50 m of primarily mafic rock is exposed in the northernmost part of the location, and the contact between the mafic and felsic rocks is extremely weathered and more densely fractured. Several intrusions, in terms of dykes and pegmatite, are abundant, but the outcrop is dominated by several E–W-trending shear zones showing an overall ductile deformation style. The shear zones are predominantly situated in the felsic gneiss, and are dominated by foliation-parallel shear, illustrated by Z-shaped drag folds and slickenlines primarily indicating dextral shear movement (Figure 4). The core of the shear zones also occasionally shows mylonites and pseudo-tachylites. Occasional thin foliation-parallel dykes occur in the southernmost part of the area, and near BBQ2 (see Figure 4) a pegmatite intrudes the banded gneiss. This pegmatite is flanked by a NE–SW-trending sinistral strike-slip fault that truncates the system 1 shear zones (and is accordingly younger). These faults are interpreted to belong to the system 5 (blue) lineaments (Figure 4).

Location C, System 1, 2 and 5 Lineaments

At the southernmost end of site area A (Figure 2), in front of the Russell Glacier, a meltwater river has eroded the front of the glacier and removed all the sediments, exposing a 200 m × 400 m outcrop including a triple junction of three lineaments: systems 1, 2 and 5. High-resolution aerial photos were acquired from the central part of the area, where the Kangerlussuaq–Russell thrust fault continues under the glacier. Deformed felsic and mafic gneisses, as well as intrusions such as the Kangâmiut dykes and pronounced pegmatites are clearly distinctive, and single faults and deformation zones are outlined (Figure 5).

The area was mapped in detail during the field season in 2013. Approximately 80 measurements were collected on fault planes, dykes and foliations, and their characteristics were described. Generally, the rock types are dominated by highly deformed banded felsic gneiss (Figure 5A) intercalated with more mafic rocks and a large variety of intruded rock types of various origins and types (Figure 5B). The most conspicuous are the 10–15-m-thick E–W-trending Kangâmiut dykes (Figure 5C) and NE–SW trending 10–50-cm-thick pegmatites (Figure 5D).

The area may be separated by distinct northern and southern deformation styles, which are separated by the central Kangerlussuaq–Russell thrust fault. The main thrust fault forms a 10–15-m-wide thrust core striking ENE–WSW and dipping 70° to NNW. The fault core consists of multiple zones with a fault gauge and strongly crushed zones, with closely spaced striated fault planes showing a dip slip reverse displacement, mixed with younger sub-horizontal dextral strike slip striations (Figure 5E). The crushing in the fault core has facilitated glacial erosion, forming a valley under the glacier. This valley can be traced in the topography more than 30 km to Kangerlussuaq. Secondary parallel thrust faults are visible in a 300-m-wide zone north and south of the fault core. Generally, the northern part of the area is less deformed than the southern part. The Kangâmiut dykes are fairly intact, in contrast to the southern area, where the gneiss is more sheared and the Kangâmiut dykes are broken up in large boudins. This is clearly a result of displacement along the Kangerlussuaq–Russell thrust fault, showing a system 1 shear zone displaced in the footwall block and the less deformed basement rocks displaced in the hanging wall block. Finally, system 5 faults striking NW–SE (120–130°) and dipping 70–60° towards SE truncate all structures, with a general sinistral displacement ranging within the order of cm to 4–5 m. The spacing of the fault planes is generally homogeneous throughout the northern area, with an average spacing of 2–10 m between the fault planes.

Location D, Kangerlussuaq Area (Site Area B)

Location D is a large area (3 km × 4 km) around Kangerlussuaq Airport marked as site area B in Figure 1. Figure 6A presents an aerial image of Kangerlussuaq Airport showing two parallel system 2 thrust faults: the Kangerlussuaq–Russell thrust fault and the corresponding Sandflugtdalen thrust fault, located approximately 1 km further south. The Sandflugtdalen thrust fault trends parallel with the airport strip, continuing east towards the Russell Glacier through Sandflugtdalen and west towards the Søndre Strømfjord. The Kangerlussuaq–Russell thrust fault forms a very distinct depression containing a number of elongated lakes that are easy to identify on aerial images, whilst the Sandflugtdalen thrust fault is situated in the middle or northern part of the Sandflugtdalen (Sand drift valley), a sediment-filled depression, which makes the exact location of the thrust fault ambiguous.

Despite the fact that the Sandflugtdalen thrust fault is difficult to detect, the significantly different degree of deformation on the N and S sides of the thrust fault indicates that a considerable amount of thrusting has occurred along the fault. The area south of the Sandflugtdalen thrust fault exhibits Kangâmiut dykes that are largely undisturbed, with a sub-vertical dip (Figure 6B–D), while the Kangâmiut dykes are strongly sheared (boudinaged) and moderately dip towards N, on the northern side of the valley, especially in the so-called Kitchen Mountain to the north (Figure 6E, F). Field investigations have demonstrated that the deformation is initially restricted to system 1 dextral semi-ductile shear zones, later truncated by the system 2 Kangerlussuaq–Russell thrust fault. A later stage of sinistral strike slip deformation occurs in some areas and N–S-trending strike-slip faulting is also observed.

The change in the deformation style of the Kangâmiut dykes is evident when observing them along a regional N–S transect from the folded area in the north (site area A) (Figure 2) and through the strongly sheared system 1 shear zones in the central area, towards the generally undisturbed area south of the Sandflugtdalen thrust fault. The folded area in the north shows Kangâmiut dykes that are mostly undisturbed or in some places locally folded during the D2 deformation phase (Figure 7), while the area with the system 1 shear zones shows boudinaged and sheared Kangâmiut dykes (Figure 6E, F).

The observations in the Kangerlussuaq area (site area B), coupled with observations further NE in site area A (Figures 1 and 2), suggest that the southernmost end of the Nagssugtoqidian orogen is complex with several geologically different belts or zones having notably different degrees of deformation. The deformation changes from folded areas (Figures 3 and 7) to steep N-dipping semi-ductile shear zones (Figure 4) and steep thrust faulting with a general dip towards NW (Figure 5), with noticeably undisturbed sections in the south (Figure 6).

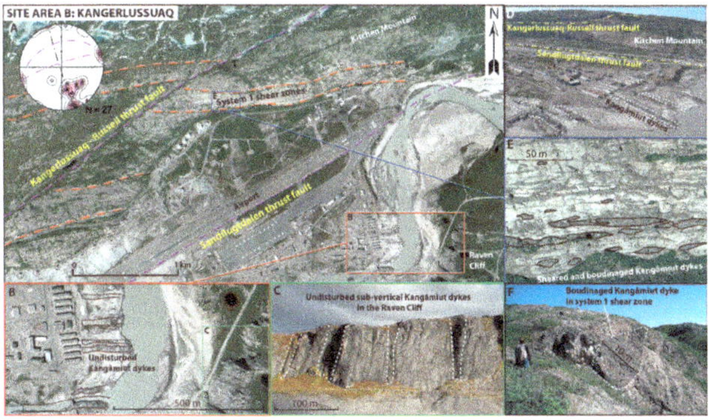

Figure 6: (**A**) Aerial image illustrating cross-cutting relationships between system 1 shear zones and system 2 thrust faulting. The Sandflugtdalen thrust fault is the major fault dividing these areas, and it is clear how the Kangerlussuaq–Russell thrust fault truncates the system 1 shear zones, displacing them by several hundreds of metres. (**B–D**) The southern part of the valley and the Raven Cliff is dominated by weakly deformed Kangâmiut dykes. (**E**, **F**) The northern part of the valley and Kitchen Mountain shows boudinaged Kangâmiut dykes deformed by the system 1 shear zones.

Figure 7: Photos showing the appearance of Kangâmiut dykes in the northern folded area of the study site. (**A**) Kangâmiut dyke intruded into the bedrock during an extensional strain regime. (**B**) A folded dyke showing ductile D2 deformation. (**C**) A large Kangâmiut dyke truncating the F1 folds and slightly folding during D2 deformation.

GEOLOGICAL EVOLUTION OF THE AREA

As demonstrated by its structural complexity, the field area has undergone several phases of deformation, and these phases have been compiled into a schematic model describing the geological evolution of the area. Two stages of ductile deformation, D1 and D2 (folding F1 and F2), have been identified. D1 deformation is considered to be the oldest discernible deformation phase in the area. The Kangâmiut dykes cross-cut the S1 foliation and the related F1 folds (Figure 8), and the intrusion of these dykes is thus a younger event. The second stage of deformation (D2) is the F2 folding phase, which consists of re-folding of the D1 folds. This folding phase also included deformation and folding of the Kangâmiut dykes at several locations, implying that this folding phase post-dates the intrusion of these dykes (Figure 8). The next deformational event comprises the E–W-trending and steeply N dipping semi-ductile shear zones (system 1). These shear zones are interpreted to be synkinematic or contemporaneous with the F2 folding event. It is possible these shear zones may have acted as a décollement surface for the folding. These shear zones also post-date the Kangâmiut dykes, since these dykes have been deformed and sheared by the shear zones (Figure 8).

The onset of the brittle deformation (D3 deformation phase) is defined by several ENE–WSW-trending thrusts or reverse faults (system 2). At the southern end of site area A, several of these thrust faults are spaced at regular intervals, building up thrust sheets with a hanging wall thrusted towards the SSE (Figure 8). These thrusts cross-cut all the aforementioned features, but they also represent a gradual rotational shift from a general strike-slip-dominated setting (passive plate boundaries) towards classic collisional structures. These thrust faults are especially prominent along the Kangerlussuaq–Russell thrust fault, with a generally moderate to steep dip towards the NW. Accordingly, these thrust sheets have been rotated towards a more steeply dipping position from the original location. The system 3, 4, 5 and 6 lineaments show a significantly more brittle character than the previously described deformation phases (Figure 8). The system 3 lineaments overprint systems 1 and 2 and the F2 folds. The normal faulting and NNE-trending pegmatites of system 4 indicate either a general extensional stress regime or local transtension during strike-slip movements. Finally, the youngest geological events are related to the type 5 and 6 lineaments, forming two generally brittle strike-slip fault systems, which are also regarded as the primary hydraulic zones in the area.

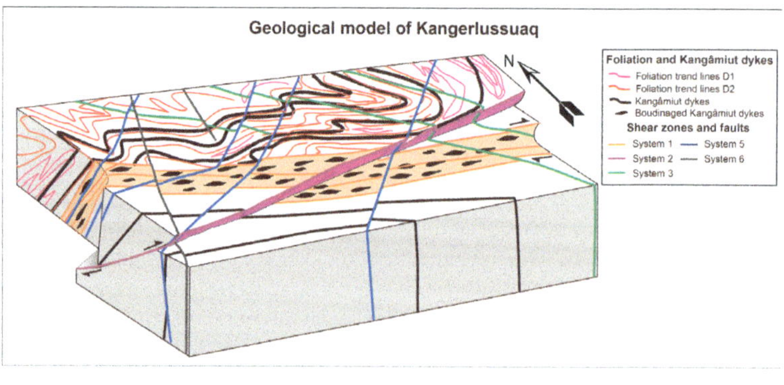

Figure 8: Schematic diagram illustrating the structural geological evolution in the Kangerlussuaq area. The first deformational phase is F1 folding, which is intruded by Kangâmiut dykes during an extensional strain regime. The next phase is the D2 ductile deformation phase, generating large-

scale NNW-plunging F2 folds, together with the syntectonic semi-ductile shear zones trending E–W and dipping towards the N, and showing a dextral sense of shear. The onset of brittle deformation is marked by thrust faults trending ENE–WSW, creating thrust sheets with a hanging wall thrusted towards the SSE. Brittle systems 3 (green, trending SE–NW), 5 (blue, trending SW–NE) and 6 (grey, trending N–S) cross-cut the aforementioned structures in different orientations.

DISCUSSION

Since no radiometric age dating has been performed in the area, the absolute ages of the different types of ductile and brittle deformations are uncertain. However, coupling with other areas within the orogen is suggested, comparing the individual types of deformation that each phase shows and recognising similar structures within the Nagssugtoqidian orogen in West Greenland (e.g., Garde and Hollis [1] and van Gool *et al.* [2] in Table 1).

Table 1: Geological history of the Kangerlussuaq area. * Based on Garde and Hollis [1] and van Gool *et al.* [2]

Age	General Tectonic Events in Central West Greenland *	Correlation to Kangerlussuaq Area
>2.5 Ga (Archaean)	Formation of banded felsic/mafic gneiss	F1 folding + multiple healed structures
ca. 2.04 Ga intrusions	Continental rifting coupled with mafic dyke intrusion (Kangâmiut mafic dykes).	Intrusion of mafic dykes (Kangâmiut mafic dykes).
1.92–1.75 Ga	Nagssugtoqidian Orogeny with continental collision. Reworking and folding of gneiss/mafic rocks followed by peak metamorphism and large scale folding during N-S contraction. Finally formation of steep belts with sinistral movements.	F2 folding event. System 1 dextral semi-ductile shear zones. System 2 WSW–ENE-trending thrust faults Kangerlussuaq-Russell and Sandflugtdalen thrust faults. System 3 SE-NW sinistral strike-slip shear zones and intrusion of pegmatites.
1.78–1.2 Ga	Various stress conditions. Formation and reactivation of open mode fractures at more shallow depths during the 20–25 km uplift.	System 4 NE-SW normal faults with shallow dip towards SE and NW accompanied by pegmatite intrusions.
1.2 Ga	Diamond-bearing ultramafic lamprophyre intrusions, S and W of Kangerlussuaq.	
600 Ma	Kimberlite intrusions S and W of Kangerlussuaq.	
100–50 Ma	Faulting related to sea-floor spreading during the opening of the Labrador Sea and Baffin Bay. NNE–SSW sinistral strike-slip faulting and conjugate dextral strike slip faulting.	System 5 sinistral strike-slip faults trending NE–SW. Reactivation of NE-SW trending pegmatites. System 6 dextral strike slip faults trending N–S.
At least the last 2 Ma	Repeated glaciations resulting in erosion, glacier induced reactivation of fractures during glacial subsidence and rebound of the basement.	Reactivation of existing fault zones.

D1 is interpreted to be Archaean in age and belongs to a deformation and metamorphic event occurring soon after the emplacement of the protolith at 2.81 and 2.72 Ga [11,12], while D2 is younger than the 2.05–2.04 Ga Kangâmiut dyke swarm [21]. This is evident because the Kangâmiut dyke swarm was folded during the D2 deformation phase. The F2 folds are interpreted to be a contemporary event with the system 1 shear zones, to be coupled with the main collision stage during the Nagssugtoqidian orogeny between 1.92 and 1.75 Ga [1]. Van Gool *et al.* [2] described isoclinal folds (F2) that deform both the large-scale tectonic contacts and S1 fabrics, forming a second generation of high-temperature structures. They concluded that the F2 folds are not consistently oriented and their kinematics are not known. Some appear directly related to thrusting, whereas others deform the thrust-related foliation, and they probably include several generations of folding. Similar shearing in high-strain, low-angle shear zones that are connected to folding has been described further north in the Norder Strømfjord area by Passchier *et al.* [31] and Manatschal *et al.* [32]. These shear zones, which both cut and are folded by folds, have a consistent top-to-the-east kinematic framework when they are unfolded back to a shallowly dipping orientation. Mineral lineations and fold axes occur in parallel. Manatschal *et al.* [32] argued that these shear zones indicate an east-vergent extensional regime based on the truncation of the Ussuit unit in some of the thrust sheets. This combination of coeval folding and shearing appears best interpreted as post-collisional north–south convergence, in which crustal thickening was compensated by an east–west escape or collapse [2].

The next phase of deformation (D3) is related to system 2 (e.g., the Kangerlussuaq–Russell thrust fault and Sandflugtdalen thrust fault), which thrust the folded and sheared rock towards SSE–SE, towards the foreland of the Atlantic craton. These thrusts have later been reactivated with a dextral strike-slip movement. Further north, the folding event is due to N–S compression, which is compatible with south-directed thrusting in the Ikertôq thrust zone approximately 30 km north of our study area. The thrusting towards SE in our area indicates a more NW–SE-directed compression, which may also indicate the strain direction causing the reactivated dextral shear sense in the more ductile 1 shear zones. The variation in deformation style in response to the same regional stress field probably reflects different levels of crustal exposure across the orogen today. The metamorphic peaks in the foreland postdate the metamorphic peak in the central parts, presumably indicating that crustal thickening was ongoing along the flanks of the Nagssugtoqidian orogen after it ceased in the central parts [2].

Kolb [8] described similar changing of the stress field for the Nagssugtoqidian orogen in Southeast Greenland, where the orientation of the stress field during later stages of collision (D2) had a NE–SW direction. This was followed by D3 deformation, with a south-eastward juxtaposition of the Switzerland Terrane, which required rotation of the regional stress field to a NW–SE orientation either at 1820 or 1740 Ma, and could well be coupled with the D3 thrusting in our area.

Accordingly, it is interpreted that the F2 folding was contemporaneous with the system 1 shear zones, and the D3 thrust zones post-date the two earlier events during the Nagssugtoqidian orogeny. The strain direction in our study area points to a NW–SE-directed contraction, which would indicate that it has been shifting since the onset of the F2 folding event, suggesting a more NE–SW-directed contraction for the folding. However, the system 1 shear zones and the system 2 thrust zones can be found in the same stress regime, simply reflecting different levels of crustal exposure indicative of the nature of their 14respective deformation. The D3 thrust zones are defined to be the youngest, as they cross-cut these earlier deformational structures.

The timing of the system 3, 4, 5 and 6 lineaments is highly speculative, however as they are more brittle, they might be related to tectonic events taking place after the orogeny. The youngest (type 6 lineaments), and hence most brittle deformations may be related to the opening of the North Atlantic Ocean, the Labrador Sea and the Baffin Bay over the last 100 Ma [30].

The reworked felsic banded gneisses and mafic garnet-rich orthogneisses found in the study area are presumed to be Archaean in age, since similar types of rocks have been described by Kalsbeek and Nutman [11] further to the west of our study area. Since no detailed petrological description or geochemistry has been performed on our rocks, their coupling to the regional context and linking of their origin to tectonic evolution in the area is not possible.

SUMMARY

The results of a detailed structural investigation of lineament zones revealed from remote sensing of geophysical and topographic data and aerial photo interpretation, together with detailed geological mapping at key locations facilitated the construction of a tectonic structural model describing the geological development of the Kangerlussuaq area. The area has undergone several episodes of deformation, which have been compiled into an event succession that recognizes eight tectonic events overprinting each other: Two stages of folding (F1 and F2) have been identified along with one major episode of intrusion of the Kangâmiut mafic dyke swarm at 2.05 Ga, and six pronounced faulting/shearing events post-dating the Kangâmiut dykes extending from ductile deformation shearing events to brittle deformation with extensive faulting, covering both the collisional and post-collisional tectonic history of the area.

Two types of lineaments are prominent in the Kangerlussuaq area; a semi-ductile type trending E–W with a dextral sense of shear and two pronounced lineaments outlining the *Kangerlussuaq–Russell thrust fault* and the *Sandflugtdalen thrust fault*.

The absolute ages of the different types of ductile and brittle deformations are uncertain because of a lack of radiometric ages. The E–W-trending semi-ductile shear zones and the Kangerlussuaq–Russell and Sandflugtdalen thrust faults are interpreted to be related to different tectonic phases during the formation of the Nagssugtoqidian orogen, together with the previously occurring F2 folding event. However, the exact relation between the F2 folding system and the system 1 shear zones remains to be determined. The system 3 and 4 semi brittle lineaments may be of intermediate ages related to later tectonic events. System 4, in particular, is related to extensional stress regimes accompanied by pegmatite intrusions. The youngest and hence most brittle deformations, systems 5 and 6, may be related to the opening of the North Atlantic Ocean, the Labrador Sea and the Baffin Bay over the last 100 Ma. An event succession model has been constructed and correlation with the regional geological event model has been attempted.

Finally, radiometric age determinations are suggested, particularly of the available felsic rocks together with cross-cutting pegmatites, and of pseudotachylites and other fault-related rocks, in order to better constrain the tectonic history. These age determinations, coupled with detailed petrological and geochemistry descriptions, would be vital to resolve the tectonic evolution of the Kangerlussuaq area, and would thus be an important addition to understanding of the geological evolution of central West Greenland.

ACKNOWLEDGMENTS

The work was funded by the Swedish, Finnish and Canadian nuclear waste management organizations and carried out by participants from GTK, GEUS, SKB, Posiva and NWMO. Special gratitude is expressed to whole GAP Subproject C field team for their contribution and endless efforts during several field seasons in the Kangerlussuaq area. The authors also wish to give a special thanks to Lillemor Claesson Liljedahl and Timo Ruskeeniemi for constructive comments on the early versions of the manuscript. The authors wish to thank both reviewers (Dr. Iain Neill and an anonymous reviewer) of the paper, who enhanced the final manuscript before publication.

REFERENCES

1. 1. Garde, A.A.; Hollis, J.A. A buried Palaeoproterozoic spreading ridge in the northern Nagssugtoqidian orogen, West Greenland. In *The Evolving Continents: Understanding Processes of Continental Growth*; Special Publications Volume 338; Kusky, T.M., Zhai, M.-G., Xiao, W., Eds.; Geological Society of London: London, UK, 2010; pp. 213–234.
2. 2. Van Gool, J.A.M.; Connelly, J.N.; Marker, M.; Mengel, F.C. The Nagssugtoqidian orogen of West Greenland: Tectonic evolution and regional correlations from a West Greenland perspective. *Can. J. Earth Sci.* 2002, *39*, 665–686.
3. 3. Ramberg, H. On the petrogenesis of the gneiss complexes between Sukkertoppen and Christianshaab, West Greenland. *Meddelelser fra Dansk Geologisk Forening* 1949, *11*, 312–327.
4. 4. Escher, A.; Pulvertaft, T.C.R. Rinkian mobile belt of West Greenland. *Geol. Greenl.* 1976, *1976*,
5. 104–119.
6. 5. Marker, M.; Mengel, F.; van Gool, J. Evolution of the Palaeoproterozoic Nagssugtoqidian orogen: DLC investigations in West Greenland. *Rapp. Grønlands. Geol. Unders.* 1995, *165*, 100–105.
7. 6. Connelly, J.N.; Thrane, K.; Krawiec, A.W.; Garde, A.A. Linking the Palaeoproterozoic Nagssugtoqidian and Rinkian orogens through the Disko Bugt region of West Greenland. *J. Geol. Soc.* 2006, *163*, 319–335.
8. 7. Harper, J.; Hubbard, A.; Ruskeeniemi, T.; Liljedahl, L.C.; Lehtinen, A.; Booth, A.; Brinkerhoff, D.; Drake, H.; Dow, C.; Doyle, S.; *et al. The Greenland Analogue Project Yearly Report 2010*; Report SKB R-11-23; Swedish Nuclear Fuel and Waste Management Co: Stockhom, Sweden, 2011.
9. 8. Kolb, J. Structure of the Palaeoproterozoic Nagssugtoqidian Orogen, South-East Greenland: Model for the tectonic evolution. *Precambrian Res.* 2014, *255*, 809–822.

10. 9. Kalsbeek, F.; Taylor, P.N.; Henriksen, N. Age of rocks, structures, and metamorphism in the Nagssugtoqidian mobile belt, West Greenland—Field and Pb-isotopic evidence. *Can. J. Earth Sci.* 1984, *21*, 1126–1131.

11. 10. Kalsbeek, F.; Pidgeon, R.T.; Taylor, P.N. Nagssugtoqidian mobile belt of West Greenland: Cryptic 1850Ma suture between two Archaean continents—Chemical and isotopic evidence. *Earth Planet. Sci. Lett.* 1987, *85*, 365–385.

12. 11. Kalsbeek, F.; Nutman, A.P. Anatomy of the early Proterozoic Nagssugtoqidian orogen, West Greenland, explored by reconnaissance SHRIMP U–Pb dating. *Geol. Soc. Am.* 1996, *24*, 515–518.

13. 12. Connelly, J.N.; Mengel, F.C. Evolution of Archaean components in the Palaeoproterozoic Nagssugtoqidian Orogen, West Greenland. *Geol. Soc. Am. Bull.* 2000, *112*, 747–763.

14. 13. Windley, B.F. Primary quartz ferro-dolerite/garnet amphibolite dikes in the Sukkertoppen region of

15. West Greenland. In Proceedings of Mechanism of Igneous Intrusion, Liverpool, UK, 9–11 January 1969; pp. 79–92.

16. 14. Escher, A.; Escher, J.C.; Watterson, J. The reorientation of the Kangâmiut dike swarm, West Greenland. *Can. J. Earth Sci.* 1975, *12*, 158–173.

17. 15. Escher, A.; Sørensen, K.; Zeck, H.P. Nagssugtoqidian Mobile Belt in West Greenland. *Geol. Greenl.* 1976, *1976*, 103.

18. 16. Korstgård, J.A. Metamorphism of the Kangâmiut dykes and the metamorphic and structural evolution of the southern Nagssugtoqidian boundary in the Itivdleq-Ikertôq region, West Greenland. *Rapp. Grønlands. Geol. Unders.* 1979, *89*, 63–75.

19. 17. Bridgwater, D.; Mengel, F.; Fryer, B.; Wagner, P.; Hansen, S.C. Early Proterozoic Mafic Dykes in the North Atlantic and Baltic Cratons: Field Setting and Chemistry of Distinctive Dyke Swarms. In *Early Precambrian Processes*; Special Publications Volume 95; Coward, M.E., Ed.; Geological Society of London: London, UK, 1995; pp. 193–210.

20. 18. Nutman, A.P.; Kalsbeek, F.; Marker, M.; van Gool, J.A.M.; Bridgwater, D. U–Pb zircon ages of Kangâmiut dykes and detrital zircons in metasediments in the Palaeoproterozoic Nagssugtoqidian Orogen (West Greenland). *Precambrian Res.* 1999, *93*, 87–104.

21. 19. Connelly, J.N.; van Gool, J.A.M.; Mengel, F.C. Temporal evolution of a deeply eroded orogen: The Nagssugtoqidian Orogen, West Greenland. *Can. J. Earth Sci.* 2000, *37*, 1121–1142.

22. 20. Cadman, A.C.; Tarney, J.; Bridgwater, D.; Mengel, F.; Whitehouse, M.J.; Windley, B.F. The petrogenesis of the Kangâmiut dyke swarm, W. Greenland. *Precambrian Res.* 2001, *105*, 183–203.

23. 21. Mayborn, K.R.; Lesher, C.E. Origin and evolution of the Kangâmiut mafic dyke swarm, West Greenland. *Geol. Surv. Den. Greenl. Bull.* 2006, *11*, 61–86.

24. 22. Taylor, P.N.; Kalsbeek, F. Dating the metamorphism of Precambrian marbles: Examples from Proterozoic mobile belts in Greenland. *Chem. Geol. Isot. Geosci. Sect.* 1990, *86*, 21–28.

25. 23. Willigers, B.J.A.; Krogstad, E.J.; Wijbrans, J.R. Comparison of thermochronometers in a slowly cooled granulite terrain: Nagssugtoqidian Orogen, West Greenland. *J. Petrol.* 2001, *42*, 1720–1749.

26. 24. Connelly, J.N.; Thrane, K. Rapid determination of Pb isotopes to define Precambrian allochthonous domains: An example from West Greenland. *Geology* 2005, *33*, 953–956.

27. 25. Garde, A.A.; Hollis, J.A.; Mazur, S. Palaeoproterozoic greenstones and pelitic schists in the northern Nagssugtoqidian orogen, West Greenland: Evidence for a second subduction zone? In Proceedings of Geological Association of Canada–Mineralogical Association of Canada (GAC–MAC) Annual Meeting, Yellowknife, Canada, 23–25 May 2007; Volume 32, p. 30.

28. 26. St.-Onge, M.; van Gool, J.A.M.; Garde, A.A.; Scott, D.J. Correlation of Archaean and Palaeoproterozoic units between north-eastern Canada and western Greenland: Constraining the pre-collisional upper plate accretionary history of the Trans-Hudson orogen. In *Earth Accretionary Systems in Space and Time*; Special Publications Volume 318; Cawood, P.A., Kröner, A., Eds.; Geological Society of London: London, UK, 2009; pp. 193–235.

29. 27. Garde, A.A.; Marker, M. *Geological Map of Greenland: Sandflugtdalen–Nuussuaq*; Geological Survey of Denmark and Greenland: Copenhagen, Denmark, 2010; Map Sheet, 1:500000.

30. 28. Klint, K.E.S.; Engström, J.; Parmenter, A.; Ruskeeneimi, T.; Liljedahl, L.C.; Lehtinen, A. Lineament mapping and geological history of the Kangerlussuaq region, southern West Greenland. *Geol. Surv. Den. Greenl. Bull.* 2013, *28*, 57–60.

31. 29. Grocott, J. Shape fabrics and superimposed simple shear strain in a Precambrian shear belt, West

32. Greenland. *J. Geol. Soc.* 1979, *136*, 471–488. *Geosciences* 2014, *4* 334

33. 30. Wilson, R.W.; Klint, K.E.S.; van Gool, J.A.M.; McCaffrey, K.J.W.; Holdsworth, R.E.; Chalmers, J.A. Faults and fractures in central West Greenland: onshore expression of continental break-up and sea-floor spreading in the Labrador—Baffin Bay Sea. *Geol. Surv. Den. Greenl. Bull.* 2006, *11*, 185–204.

34. 31. Passchier, C.W.; den Brok, S.W.J.; van Gool, J.A.M.; Marker, M.; Manatschal, G. A laterally constricted shear zone system—The Nordre Strømfjord steep belt, Nagssugtoqidian Orogen, W. Greenland. *Terra Nova* 1997, *9*, 199–202.

35. 32. Manatschal, G.; Ulfbeck, D.; van Gool, J.A.M. Change from thrusting to syn-orogenic extension in

36. a mid-crustal level: An example from the Ussuit area, Nagssugtoqidian Orogen, West Greenland. *Can. J. Earth Sci.* 1998, *35*, 802–819.

6
CHAPTER

UAV THERMAL INFRARED REMOTE SENSING

Stefania Amici, Matteo Turci, Salvatore Giammanco, Letizia Spampinato and Fabrizio Giulietti

Keywords: RPAS; Mud Volcano; Thermal Map

INTRODUCTION

Volcanic activity is a globally important phenomenon, affecting both the environment and human life. Satellite and airborne remote sensing methodologies have proven to be extremely valuable for examining and evaluating the effects of eruptions over a large scale. High temporal resolution systems such as MODIS (Moderate Resolution Imaging Spectroradiometer) have been used for mapping the evolution in time of volcanic plumes [1-3]. High spatial resolution systems such as ASTER (Advanced Spaceborne Thermal Emission and Reflection Radiometer) have been extensively used for global scale mapping of volcanoes [4]. During an eruptive crisis, on demand, airborne acquisition has provided details on the phenomena, allowing for the description of the morphology and evolution of lava flow fields [5,6], as well as thermal anomalies characterisation [7].

In studies of precursory signals, major indicators of upcoming eruptions include: volcanic tremors [8,9], ground deformation [10], thermal anomalies and gas emissions [11]. Seismic tremor and volcanic gas emissions, for example, are related to upcoming eruptive activity, however the mechanisms are not yet completely understood and more data, as well as new parameters, are needed. Surface temperature anomalies have been remotely investigated, however their use has been limited so far by the poor spatiotemporal resolution of satellite systems or by the expensive costs of dedicated airborne measurements.

Citation: S. Amici, M. Turci, S. Giammanco, L. Spampinato and F. Giulietti, "UAV Thermal Infrared Remote Sensing," Advances in Remote Sensing, Vol. 2 No. 4, 2013, pp. 358-364. doi:10.4236/ars.2013.24038.

Furthermore, this type of volcanic activity rarely maintains a stable level of activity, which is why reliable forecasts should be made by well-organized volcano observatories that operate at a local scale. Italy has many active volcanoes that are continuously monitored by the networks of instruments held by the IstitutoNazionale di Geofisica e Vulcanologia (INGV), which operates at both ground and airborne scale.

INGV has been using hand-held thermal cameras since 2001 during both ground-based and helicopter operated surveys for the surveillance and monitoring of Mt. Etna, Stromboli, Vulcano, Campi Flegrei and Vesuvius [3,12-17]. Indeed, airborne surveys provide information needed to get an overall view of the eruptive scenarios, particularly in the case of wide active areas such as lava flow fields, crater fumaroles and hydrothermal zones [18]. These surveys provide information to be used for thermal imaging-derived parameters, such as heat budgets and effusion rates [13,17,19]. However, one of the main drawbacks in the use of helicopters is their cost and the risk for operators and pilots, especially during volcanic paroxysmal phases.

In order to find a compromise between the advantages and disadvantages of air surveys, in 2004 INGV started to investigate the possibility to use a remotely operated Piloted Air System (RPAS) for volcanic environment applications. The first test was carried out by a joint collaboration between INGV and Università di Bologna at Stromboli volcano [20,21]. Since the first test, the activity has evolved with the aim of equipping systems with sophisticated instrumentation. Great efforts have been made in order to study how to improve technical systems and instruments for risk management within the area of volcanology. Telemetry and, in real time, acquisition systems, have been developed and tested at laboratory and field scales [22]. In this paper, we describe a field experiment performed in June 2012 involving a thermal camera equipped on a hexacopter over the Salinelle mud volcano (Sicily, Italy). Firstly, we describe the main platform and the characteristics of the sensor used for the experiment; secondly we describe the geological setting of the investigated area. Finally, we discuss the results, including the cross comparison between in-flight and *in-situ* independent thermal mapping. Specifically, we not only derived a qualitative hot-cold map of the mud volcano, but also associated the temperature values.

METHODS

The Multirotor System

The unmanned system used for our purposes is a rotor- craft with six rotors (hexacopter). This kind of aircraft uses fixed-pitch blades and a brushless motor rigidly fixed to the structure. As the speed of each rotor can be independently regulated, the control of the model, in terms of attitude, position and movement, is obtained by modifying the trust and torque produced by each rotor; this is obtained simply by modifying the relative speed of each rotor.

A Flight Controller Board (FCB), which includes a GPS receiver, air pressure sensor and Inertial Measurement Unit, is installed to aid with any instability of the system. The FCB allows for the use the of radio-controlled system.

The hexacopter used in our work is the Easy fly A2500 (www.salengeneering.it); this model has enhanced rigidity and weight via its use of carbon fiber parts instead of aluminium. It has a diameter of 100 cm and a height from ground to centre plate of 30 cm. It is powered by two battery packs (5000 + 5000 mAh) allowing an autonomous flight of approximately 15 - 20 minutes according to the batteries and payload (Figure 1). In our configuration, the payload weight used was 1.7 kg and the estimated maximum ceiling of 200 m (70 m by ENAC Italian regulation). Due to limitations in national regulations relating to RPAS, the system flew within line of sight, by radio control. For the experiment, the hexacopter was equipped with a thermal EYE 3600 AS (TE 3600 AS) designed for Original Equipment Manufacturers (OEMs). The TE 3600 AS is based on microbolometer technology, consisting of 30 μm pitch detectors, allowing for its light weight of 67 g. The spectral response is within the range 7 - 14 μm, with a thermal sensitivity less than 50 mK and a saturation temperature of 600°C. The camera has a frame rate of 25 hertz real time and generates a PAL video output. The camera is characterized by an automatic contrast that is decreased and increased according to the scene. For each thermogram acquired, a calibrated temperature value is provided corresponding to the central spot, while a colour scale bar within the image shows the minimum and maximum temperature range, expressed in decimal Celsius degrees. To integrate the camera with the system, an acquisition system has been developed using commercial components. The result is a cubic embedded solution (12 cm × 11 cm × 8 cm, and weight of 600 gr) composed of four PC-104 compliant stackable modules: Power supply, Frame grabber, PC and Wi-Fi module [22].

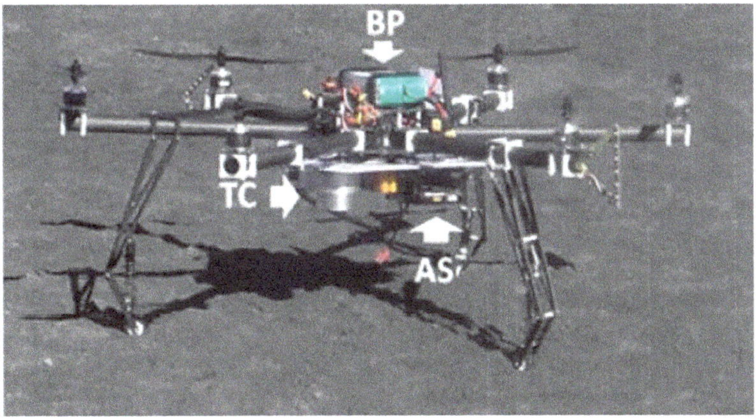

Figure 1. The hexacopter UAV system used in this study: the locations of battery pack (BP), thermal camera (TC) and acquisition system (AS) are shown.

In order to fix both the camera and the acquisition system on the hexacopter, MDV designed and realized a specific set-up using aluminium due to its light weight. A standard female USB connector allows for the GUI communication and a DB9 connector reports all camera pins. An analogue video output was configured within the system. The final flight configuration, including payload and acquisition system is shown in Figure 1.

Geological Setting of the Salinelle Mud Volcano

The Salinelle mud volcanoes occur in three sites located on the lower SW flank of Mt. Etna volcano (Italy), Figure 2. The geology of the area includes: 1) volcanic products erupted from Mt. Etna (Mongibello cycle, dated 35 ka to present, [23] that outcrop in the northern and north- eastern parts of the area; 2) terraced alluvial deposits (mostly conglomerates and clays) dated to 125 to 40 ka [23]; 3) small outcrops of alkali lavas belonging to the Timpe phase [24] that are present in the central and north-western parts of the area, and are best represented by the volcanic neck of Paternò [25], whose age was calculated at 168 ka by [26] or at 210 ka by Condomines [27]; 4) sub-alkalic lavas dated to 580 to 250 ka [23]; 5) Early-Middle Pleistocene foredeep deposits (1200 - 600 ka) mostly made of marly clays [23] that fill the Catania basin and outcrop in the eastern part of the area; 6) units of the Apenninic-Maghrebian collision belt, mostly composed of quartz-arenites and clays [28,23]. For the purposes of this study, the best site for the test was that of the Salinelle of Paternò Stadium, located at the NW boundary of the city of Paternò, very close to the local football stadium. At this site, fluid emissions occur on alkali lavas close to the Paternò volcanic neck (Figure 2(c)). The fluids escaping from these manifestations are composed of a gas phase and hypersaline waters that often form pools that range in size from centimetres to meters in diameter.

Figure 2. (a) Visible image as acquired by GeoEYE (Google Earth, 37°21'64"N 14°53'25.12"). Inset (b) shows the area selected for take-off and landing of the hexacopter UAV system; inset (c) shows the location of the three mud volcanoes named "Le Salinelle".

The mud and water mixtures are highly variable. In some cases, mud is the only fluid erupted along with gas, and builds cones up to a few meters high, with a base diameter up to ten meters. Emitted waters have a uniform and constant chemical compositions and physical-chemical characteristics; with a pH of approximately 6, an electrical conductivity of approximately 88 mS/cm, the chemical abundance of major species in the solution (on equivalent basis) is Na >> Ca > Mg > K and Cl^- >> HCO^-_3 > SO^4 = [29-31]. During the summer, intense evaporation produces large salt crusts around the emission vents, hence the name of such manifestations (from the Italian word "sale", which means salt). Most of the pools and mud vents are ephemeral and show strong temporal variations in the gas flow. Water outlet temperatures are normally between 10°C and 20°C, but occasionally increase to 40°C - 50°C [31,32]. Such anomalous temperature values are always accompanied by increases in the gas efflux.

Many paroxysms occurred in the degassing activity of the Salinelle in the past. Reports of strong eruptive activities in early 1866 and late 1878, with fountains of muddy water of up to 3 m high and water temperatures of up to 46°C, that the author relates to local seismic events that occurred some days/weeks prior to the gas eruptions [31,32]. During the past 15 years, similar increases in degassing activity have also been recorded at the Salinelle [S. Giammanco, unpublished data]. These events normally lasted from two to four months. The main component of the gas escaping from the study sites is CO_2. In all of the sites, air components are very low or are absent. Methane is usually the second most abundant gas species, although with highly variable amounts among the sites. Other minor species emitted include N_2, He, H_2, CO, H_2S, Rn and other light hydrocarbons [29,30].

Normally, gases from the Salinelle are sampled using inverted funnels placed on the ground over the smallest vents, or on the water/mud surface right over gas bubbling areas in the case of larger vents, where accessible. Funnels are connected to gas flasks for gas collection. Field measurements of water/mud temperature are normally carried out using thermocouples when vents are directly accessible or using laser devices from a distance. Since 2006, water/mud temperature measurements were carried out only using a non-contact laser-guided IR temperature meter (Fluke Corporation, WA, US, mod. Fluke 68), with accuracy of ±1% for targets above 23°C and of ±2°C for targets between −18°C and 23°C. However, due to the large variability both in the position of gas and water vents and in the gas flux, the use of devices able to easily and quickly monitor the whole degassing area is fundamental.

Experimental Setup and Results

The experimentation was carried out on 26 June 2012. The hexacopter was pre-assembled while payload and system were configured *in situ*. The weather conditions were favourable though the day: the air was extremely hot even in the afternoon (up to 32.2°C, as reported by meteo Sicilia, http://meteosicilia.it). The flights were planned during the early evening in order to reduce the effects of heat and reflection from the ground [33]. The autonomy of the system was about 15 minutes; therefore, it was only possible to carry out two flights to cover two different areas of the Salinelle. Although the site has a good accessibility, it is located in an urban area, hence both take-off and landing had to be performed at the edge of the mud volcano area, a zone characterized by an unstable muddy surface (Figure 3). In order to verify the suitability of the method in extreme environment conditions, we tested the stability of the acquisition and of the transmission systems. Then we investigated the possibility to obtain both qualitative (mapping of the active degassing areas) and quantitative (temperature distribution) thermal information. In order to obtain a quantitative mapping of the Salinelle volcano, a Forward Looking InfraRed (FLIR) thermal camera was used. The FLIR camera used was a A310 model, consisting of a 320 × 240 microbolometer detector array operating in the 8-14 μm waveband, with dynamic range of 0°C to 350°C and accuracy of readings of ±2% (see http://www.flir.com for more details).

Six thermograms from the video recorded by thermal EYE 3600 in correspondence to six areas in the FLIR image were processed [34] in order to choose an area for comparison, Figure 4.

Similar to the ground-based temperature measurements of a cold pool yielding mean values of 34.7°C (Figure 4(b)), apparent temperatures of the same pool measured in-flight gave values of 34°C (Figure 5(c)). The strong agreement was reached despite the different viewing geometry of the two cameras (*i.e.* horizontal in the case of the FLIR camera, vertical in the case of the drone-borne camera).

An emissivity value of 0.95 was used, derived using an emissivity spectrum measured by μFTIR one day after the flights (Figure 5), in agreement with the same values for soil saturated with water [35].

The second flight was performed in the area that is usually mapped by a field laser system. This site is characterized by a large area affected by water and CO_2 emissions. Figure 6 shows the area investigated during the second hexacopter flight and the thermal frame acquired in correspondence of the site regularly mapped. The temperature detected at the centre of the TC3600 image is 22°C.

Figure 7 shows the time series of the Salinelle highest temperature acquired during direct field surveys carried out from 2009 to 2012.

Figure 3. The frame composite visible image shows the reconstructed pattern of the first hexacopter flight.

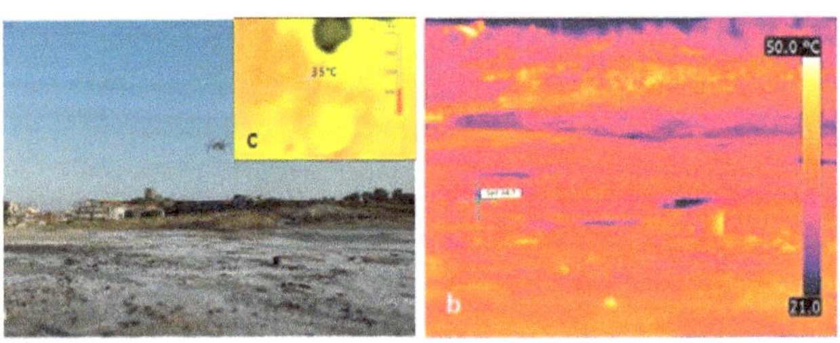

Figure 4. (a) Visible image of the area selected for the first flight; (b) thermal image acquired by FLIR camera, the spot is located in correspondence of the same area as seen by the TC 3600 (c).

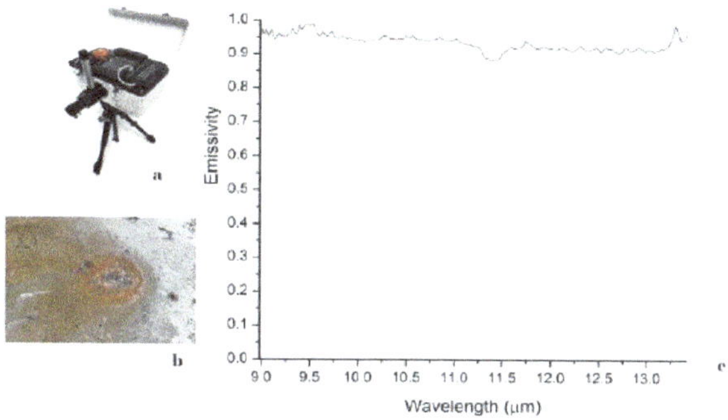

Figure 5. (a) **μFTIR** instrument used to measure emissivity spectrum; (b) Visible image of cold pool (20 cm diameter); (c) Measured emissivity spectrum.

Figure 6. The frame composite visible image shows the reconstructed pattern of the second hexacopter flight. Inset shows the frame from the thermal camera mounted on the drone taken on the same site as field temperature measurements and the temperature measured in the centre of the image.

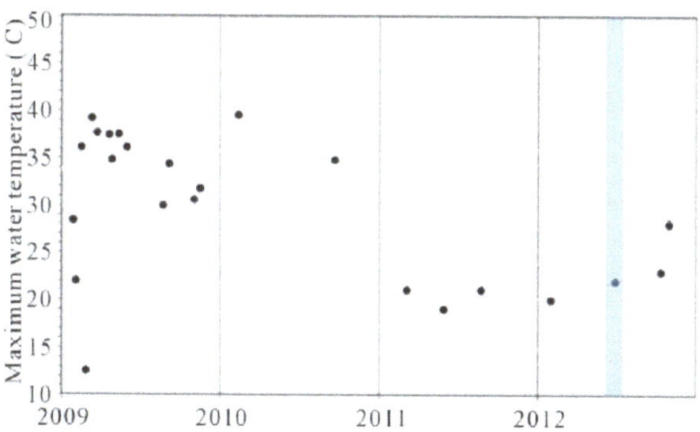

Figure 7. Temporal pattern of the highest water temperature measured in the mud pools of the Salinelle mud volcanoes during the period 2009-2012 using field instruments. The vertical blue

stripe highlights the temperature measured on June 27, 2012. A thermogram corresponding to the same site as this field measurement has been extracted and shown in figure 6.

The vertical blue stripe outlines the measurement acquired on June 27, 2012, during the period of time when the hexacopter flights were performed. The measured temperature was 22°C ± 0.4°C, therefore comparable to the temperature value extracted from the frame measured with the thermal camera mounted on the drone (see Figure 6).

CONCLUSIONS

Thermal mapping of volcanoes at super-spatial resolution scale is important for investigating thermal anomalies that can relate to upcoming volcanic eruptions. The RPAS cutting edge technology and a sophisticated lightweight thermal camera have been used in an experiment conducted on a very active Italian mud volcano: the Salinelle of Paternò. The mud volcano is located on the southwest boundary of Mt. Etna volcano (Italy).

The site was chosen for the flight test for several reasons: relatively easy access and peculiar geological features which in certain case are related to Etna activity. Furthermore, it is an example of an area difficult to be monitored using *in situ* measurements due to the muddy surface.

A hexacopter thermally equipped and connected with an embedded acquisition system specifically designed for this application was flown above the mud volcano. The experiment was performed in June 2012 and two areas were mapped: an area never mapped before and an area that is usually mapped by *in situ* measurements. This second area was used to cross validate images acquired by the hexacopter with images acquired by a ground FLIR camera. The thermal camera flew above the usually measured area and the acquired temperature was compared to the temperature obtained by a laser measurement time series (2009-2012) showing a good correspondence. The experiment was a success from both a technical and scientific point of view. Technically speaking, the flying system worked appropriately; a thermal video was recorded and transmitted in real time during the whole flight time, despite take-off and landing occurring under extreme environmental conditions. Scientifically speaking, we found a strong agreement between temperatures acquired in-flight and those acquired *in-situ* with different field instruments. Thermal imaging on board an unmanned system therefore proves to be a promising technique for distinguish thermal gradient and temperature values for volcanic applications. In particular, we would like to use the hexacopter over some areas of Mount Etna, which exhibit thermal anomalies and are very hard to be measured with high resolution.

ACKNOWLEDGEMENTS

We would like to thank Dr. A. La Spina, and T. Caltabiano of Catania INGV branch for logistic and support in field measurements. A special thanks to Dr M. F. Buon- giorno head of CNT Unit of Roma INGV for supporting this research. The authors wish to thank Mario Gattelli and Massimiliano Pompignoli for providing the experi- mental UAV airframe.

REFERENCES

1. R. Wright, L. L. Flynn, H. Garbeil, A. Harris and E. Pilger, "Automated Volcanic Eruption Detection Using MODIS," Remote Sensing of Environment, Vol. 82, No. 1, 2002, pp. 135-155. http://dx.doi.org/10.1016/S0034-4257(02)00030-5

2. S. Corradini, C. Spinetti, E. Carboni, C. Tirelli, M. F. Buongiorno, et al., "Mt. Etna Tropospheric Ash Retrieval and Sensitivity Analysis Using Moderate Resolution Imaging Spectroradiometer Measurements," Journal of Applied Remote Sensing, Vol. 2, No. 1, 2008, Article ID: 023550. http://dx.doi.org/10.1117/1.3046674

3. S. Corradini, L. Merucci, A. J. Prata and A. Piscini, "Volcanic Ash and SO_2 **in the 2008** Kasatochi Eruption: Retrievals Comparison from Different IR Satellite Sensors," Journal of Geophysical Research Atmospheres, Vol. 115, No. D2, 2010, pp. 2156-2202.http://dx.doi.org/10.1029/2009JD013634

4. D. Pieri and M. Abrams, "ASTER Watches the World's Volcanoes: A New Paradigm for Volcanological Observations from Orbit," Journal of Volcanology and Geothermal Research, Vol. 135, No. 1-2, 2004, pp. 13-28.http://dx.doi.org/10.1016/j.jvolgeores.2003.12.018

5. V. Lombardo, M. F. Buongiorno, D. Pieri and L. Merucci, "Differences in Landsat TM Derived Lava Flow Thermal Structures during Summit and Flank Eruption at Mount Etna," Journal of Volcanology and Geothermal Research, Vol. 134, No. 1-2, 2004, pp. 15-34.http://dx.doi.org/10.1016/j.jvolgeores.2003.12.006

6. C. Spinetti, V. Carrère, M. F. Buongiorno, A. J. Sutton and T. Elia, "Carbon Dioxide of Pu`u`O`o Volcanic Plume at Kilauea Retrieved by AVIRIS Hyperspectral Data," Remote Sensing of Environment, Vol. 112, No. 6, 2008, pp. 3192-3199. http://dx.doi.org/10.1016/j.rse.2008.03.010

7. D. Pieri and M. Abrams, "ASTER Observations of Thermal Anomalies Preceding the April 2003 Eruption of Chikurachki Volcano, Kurile Islands, Russia," Remote Sensing of Environment, Vol. 99, No. 1-2, 2005, pp. 84-94.http://dx.doi.org/10.1016/j.rse.2005.06.012

8. R. Schick, "Volcanic Tremor-Source Mechanisms and Correlation with Eruptive Activity," Natural Hazards, Vol. 1, No. 2, 1988, pp. 125-144.http://dx.doi.org/10.1007/BF00126610

9. M. Cosentino, G. Lombardo and E. Privitera, "A Model for Internal Dynamical Processes on Mt Etna," Geophysical Journal International, Vol. 97, No. 3, 1989, pp. 367-379.http://dx.doi.org/10.1111/j.1365-246X.1989.tb00508.x

10. W. J. Glyn and R. Hazel, "Detecting Volcanic Eruption Precursors: A New Method Using Gravity and Deformation Measurements," Journal of Volcanology and Geothermal Research, Vol. 113, No. 3-4, 2002, pp. 379-389. http://dx.doi.org/10.1016/S0377-0273(01)00272-4

11. A. Aiuppa, R. Moretti, C. Federico, G. Giudice, S. Gurrieri, M. Liuzzo, P. Papale, H. Shinohara and M. Valenza, "Forecasting Etna Eruptions by Real-Time Observation of Volcanic Gas Composition," Geology, Vol. 35, No. 12, 2007, pp. 1115-1118.http://dx.doi.org/10.1130/G24149A.1

12. V. Lombardo and M. F. Buongiorno, "Temperature Distribution Analysis of July 2001 Mt. Etna Eruption Observed by the Airborne Hyperspectral Sensor MIVIS," Annals of Geophysics, Vol. 46, No. 6, 2003, pp. 1217- 1228.

13. S. Calvari, L. Spampinato, L. Lodato, A. J. L. Harris, M. R. Patrick, J. Dehn, M. R. Burton and D. Andronico, "Chronology and Complex Volcanic Processes during the 2002-2003 Flank Eruption at Stromboli Volcano (Italy) Reconstructed from Direct Observations and Surveys with a Handheld Thermal Camera," Journal of Geophysical Research: Solid Earth, Vol. 110, No. B2, 2005, Article ID: B02201. http://dx.doi.org/10.1029/2004JB003129

14. G. Chiodini, G. Vilardo, V. Augusti, D. Granieri, S. Caliro, C. Minopoli and C. Terranova, "Thermal Monitoring of Hydrothermal Activity by Permanent Infrared Automatic Stations: Results Obtained at Solfatara di Pozzuoli, Campi Flegrei (Italy)," Journal of Geophysical Research: Solid Earth, Vol. 112, No. B12, 2007, Article ID: B12206.http://dx.doi.org/10.1029/2007JB005140

15. C. Spinetti and M. F. Buongiorno, "Volcanic Aerosol Optical Characteristics of Mt. Etna Tropospheric Plume Retrieved by Means of Airborne Multispectral Images," Journal of Atmospheric and Solar-Terrestrial Physics, Vol. 69, No. 9, 2007, pp. 981-994.

16. L. Spampinato, S. Calvari, C. Oppenheimer and L. Lodato, "Shallow Magma Transport for the 2002-3 Mt. Etna Eruption Inferred from Thermal Infrared Surveys," Journal of Volcanology and Geothermal Research, Vol. 177, No. 2, 2008, pp. 301-312.http://dx.doi.org/10.1016/j.jvolgeores.2008.05.013

17. A. J. L. Harris, L. Lodato, J. Dehn and L. Spampinato, "Thermal Characterization of the Vulcano Fumarole Field," Bulletin of Volcanology, Vol. 71, No. 4, 2009, pp. 441-458.http://dx.doi.org/10.1007/s00445-008-0236-8

18. P. Allard, M. Burton and F. Muré, "Spectroscopic Evidence for a Lava Fountain Driven by Previously Accumulated Magmatic Gas," Nature Letters, Vol. 433, No. 7024, 2005, pp. 407-410.

19. L. Spampinato, S. Calvari, C. Oppenheimer and E. Boschi, "Volcano Surveillance Using Infrared Cameras," Earth-Science Reviews, Vol. 106, No. 1-2, 2011, pp. 63- 91.http://dx.doi.org/10.1016/j.earscirev.2011.01.003

20. S. Amici, G. M. Saggiani, M. F Buongiorno, F. Persiani, A. Ceruti, P. Tortora, E. Troiani, F. Giulietti, G. DiStefano, D. Pieri, G. G. Bentini M. Bianconi, A. Cerutti, A. Nubile, S. Sugliani, S. Chiarini, G. Pennestrì and S. Petrini, "A UAV System for Observing Volcanoes and Natural Hazards," Fall Meeting of American Geophysical Union, San Francisco, 11-14 December 2007.

21. G. M. Saggiani, F. Persiani, A. Ceruti, P. Tortora, E. Troiani, F. Giulietti, S. Amici, M. F. Buongiorno, G. G. Bentini, M. Bianconi, A. Cerutti, A. Nubile, S. Sugliani, M. Chiarini, G. Pennestrì, S. Petrini and R. Guzzi, "UAV System Development for the Monitoring and Study Volcanic and Natural Hazard Events," Proceeding of RSPSOC, New Castle, 11-14 September 2007.

22. M. Turci, S. Amici, M. F. Buongiorno, F. Giulietti and N. A. Melega, "A Thermal EYE on Unmanned Aircraft System," Technical Report INGV, 2011, ISSN-2039-7941, n. 210.

23. S. Catalano, S. Torrisi and C. Ferlito, "The Relationship between Late Quaternary Deformation and Volcanism of Mt. Etna (eastern Sicily): New Evidence from the Sedimentary Substratum in the Catania Region," Journal of Volcanology and Geothermal Research, Vol. 132, No. 4, 2004, pp. 311-334. http://dx.doi.org/10.1016/S0377-0273(03)00433-5 [Paper reference:4]

24. S. Branca, M. Coltelli and G. Groppelli, "Geological Evolution of Etna Volcano," In: A. Bonaccorso, S. Calvari, M. Coltelli, C. Del Negro and S. Falsaperla, Eds., Mt. Etna: Volcano Laboratory, American Geophysical Union, Washington DC, 2004, pp. 49-63.

25. D. K. Chester, A. M. Duncan, J. E. Guest and C. R. J. Kilburn, "Mount Etna: The Anatomy of a Volcano," Chapman and Hall, London, 1985.

26. P. Y. Gillot, G. Kieffer and R. Romano, "The Evolution of Mount Etna in the Light of Potassium-Argon Dating," Acta Vulcanologica, Vol. 5, 1994, pp. 81-87.

27. M. Condomines, J. Tanguy and V. Michaud, "Magma Dynamics at Mt. Etna—Constraints from U-Th-Ra-Pb Radioactive Disequilibria and Sr Isotopes in Historical Lavas," Earth and Planetary Science Letters, Vol. 132, No. 1-4, 1995, pp. 25-41.http://dx.doi.org/10.1016/0012-821X(95)00052-E

28. F. Lentini, "The Geology of Mt. Etna Bas," Mem. Soc. Geol. It., Vol. 23, 1982, pp. 7G-25G.

29. G. Chiodini, W. D'Alessandro and F. Parello, "1982. Geochemistry of Gases and Waters Discharged by the Mud Volcanoes at Paternò, Mt. Etna (Italy)," Bulletin of Volcanology, Vol. 58, No. 1, 1996, pp. 51-58. http://dx.doi.org/10.1007/s004450050125

30. W. D'Alessandro, F. Parello and M. Valenza, "Gas Manifestations of Mount Etna Area: Historical Notices and New Geochemical Data (1990-1993)," Acta Vulcanologia, Vol. 8, No. 1, 1996, pp. 23-29.

31. O. Silvestri, "Le Salse e la Eruzione di Fango di Paternò (Sicilia)," Osservazioni e Ricerche. Stabilimento Tipografico, C. Galatola, Catania, 1866, p. 30.

32. O. Silvestri, "Importante Eruzione di Fango Comparsa a Paternò Nelle Adiacenze Dell'Etna ai Primi di Dicembre 1878," Bollettino di Vulcanologia Italiana, Vol. 5, 1878, pp. 131-132.

33. L. Spampinato, C. Oppenheimer, A. Cannata, P. Montalto, G. G. Salerno and S. Calvari, "On the Time-Scale of Thermal Cycles Associated with Open-Vent Degassing," Bulletin of Volcanology, Vol. 74, No. 6, 2012, pp. 1281- 1292. http://dx.doi.org/10.1007/s00445-012-0592-2

34. S. Amici, M. Turci. F. Giulietti, S. Giammanco, M. F. Buongiorno, A. La Spina and L. Spampinato, "Volcanic Environments Monitoring by Drones Mud Volcano Case Study," International Archives of the Photogrammetry, Remote Sensing and Spatial Information Sciences, Vol. XL-1/W2, UAV-G, 4-6-2013, Rostock, 2013, p. 1.

35. http://www.infrared-thermography.com/material-1.htm

7

CHAPTER

INTEGRATED ASSESSMENT OF FOREST COVER CHANGE AND ABOVE-GROUND CARBON STOCK IN PUGU AND KAZIMZUMBWI FOREST RESERVES, TANZANIA

Japhet J. Kashaigili, Makarius V. Mdemu, Augustino R. Nduganda, Boniface P. Mbilinyi

Keywords: Land Cover Change; Remote Sensing and GIS; Pugu & Kazimzumbwi Forest Reserves; Carbon Stock; Coastal Forests; Tanzania

INTRODUCTION

Global warming, which is mainly a result of increased anthropogenic Green House Gases (GHGs), represents a significant development challenge for the 21st Century. It is well known that the main cause of global warming is the increase of anthropogenic carbon dioxide (CO_2) emission, which accounts to 77% of all GHGs [1].

Reducing carbon emissions is of great important in this era of climate change. Several mechanisms have been developed by the United Nations Framework Convention on Climate Change (UNFCCC), these includes cutting down CO_2 emissions from Annex 1 countries, and reducing deforestation and forest degradation in developing countries. Although UNFCCC requires the reduction of CO_2 emissions, tropical deforestation contributes to 20-25% of annual global emissions of CO_2 [2] and it is held to be the second largest global source of anthropogenic CO_2 emissions [3].

The Reducing Emissions from Deforestation and forest Degradation (REDD) concept requires developing countries to be involved in the management of forests, including reforestation and afforestation. The REDD concept is proposing to finance all activities that contributes to the improvement of forest conditions, whereby developing

Citation: Japhet J. Kashaigili, Makarius V. Mdemu, Augustino R. Nduganda, Boniface P. Mbilinyi; Integrated Assessment of Forest Cover Change and Above-Ground Carbon Stock in Pugu and Kazimzumbwi Forest Reserves, Tanzania, Advances in Remote Sensing, 2013, 2, 1-9, http://dx.doi.org/10.4236/ars.2013.21001

countries that contribute to the reduction of carbon emissions may be able to sell carbon credits to the international carbon markets. In order to finance for carbon emission reduction, the quantification of forest carbon is of great important.

This paper presents an attempt to integrate satellite imagery and ground based inventories for an estimation of above ground tree biomass and carbon for the Pugu and Kazimzumbwi Forest Reserves (PKFR) in the coastal areas of Tanzania. PKFR are amongst the important biodiversity hotspots in Tropical Africa [4], however they are under threat of being depleted due to anthropogenic activities, including deforestation and forest degradation. Due to these anthropogenic activities, it is not well known how much forest cover has changed and the quantities of the available above ground tree carbon stocks are not known.

MATERIALS AND METHODS

Study Area

The Pugu and Kazimzumbwi Forest Reserves (PKFR) are located in the Kisarawe District of the Coastal region of **Tanzania (Figure 1).** The Kisarawe District is located at 7°10'0S, 38°49'60E and bordered by the following districts: Ilala to the East, Kibaha to the North, Morogoro to the West and Rufiji to the South. The Kisarawe District receives an annual rainfall of approximately 1236 mm, which is 20% more than Dar es Salaam. The area experiences two rainfall regimes. It receives short rains (*Vuli*) from October to December and the long rains (*Masika*) from March to May. Temperature in the District ranges between 24-31°C [5]. According to the 2002 Census, the population of Kisarawe District was 95,615 [6]. Currently, the population has continued to increase as a result of human birth and immigration. The PKFR are adjacent to each other (Figure 1), one to the north and the other one to the south. The altitude ranges between 100 and 350 m above mean sea level for PFR and 120 and 270 m above mean sea level for KFR [5].

METHODS

Remote Sensing Data and Its Processing

Table 1 presents the Landsat images used for the study. All images were acquired during the dry season, between June and July, so as to minimize seasonality and cloud effects.

Table 1: Remote sensed data used in the analysis of forest cover changes

Image	Path/Row	Acquisition date	Season
Landsat MSS	166/65	27/07/1980	Dry
Landsat TM	166/65	09/06/1995	Dry
Landsat TM	166/65	07/07/2010	Dry

TM = Thematic Mapper; MSS = Multi Spectral Scanner.

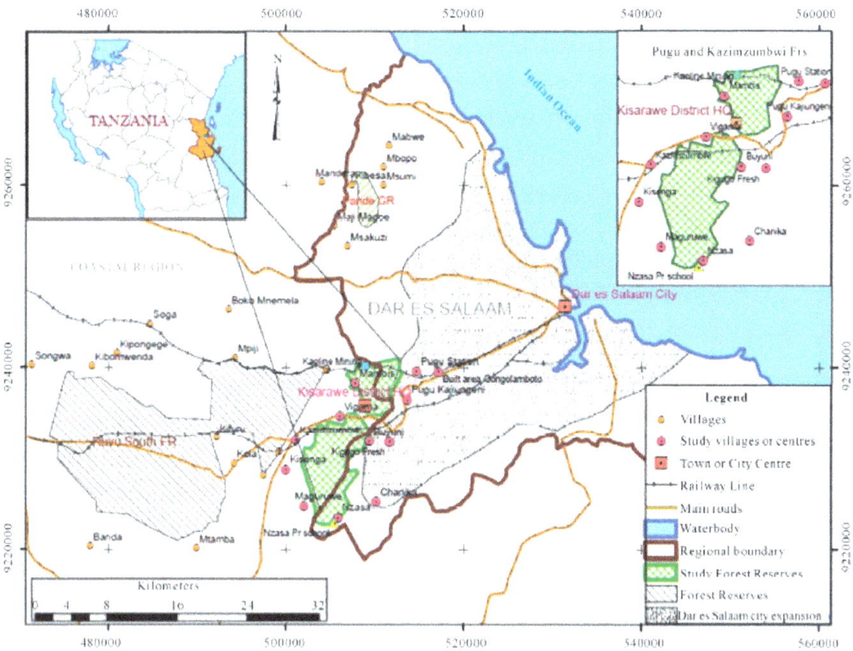

Figure 1: Location of PKFR in Kisarawe District.

The acquired image scenes of the years 1980 and 2010 had already been geo-rectified by the supplier. To ensure accurate identification of temporal changes and geometric compatibility with other sources of information, image to image geo-correction was conducted to rectify the 1995 imagery based on the 2010 image. Images enhancement was performed using a 4,5,3 colour composite band combination and its contrast was stretched using the Gaussian distribution function, followed by a high pass filter of 3 × 3 to increase the visibility of the ground control points in both images. A first order polynomial transformation and nearest neighbourhood interpolation [7] was applied to geometrically rectified the 1995 imagery and subsequently register it to the UTM map coordinate system, Zone 37 South, Datum Arc 1960.

Base maps were prepared based on the image acquired on 7th July 2010 and used in a ground truthing exercise. The essence of conducting ground truthing was to verify different cover types as described on the base maps, and for the collection of ground points for the classification accuracy assessment. Supervised classification, using the Maximum Likelihood Classifier [7-9] was performed with the ERDAS IMAGINE software. Training fields were identified by inspecting an enhanced colour composite imagery. Areas with similar spectral characteristics were trained and classified. The error matrices [10] were prepared and used in the computation of the Kappa coefficient for the classification accuracy assessment.

To analyse the changes between different time epochs, change detection analysis was performed. Many change detection methods have been developed and used for various applications. However, they can broadly be divided into: post-classification approaches and spectral change detection approaches [8]. The post classification change detection method was applied, followed by spatial overlay analysis [11] using ArcGIS, resulting in attribute tables. The tables were exported to MS-Excel to compile an area change detection matrix for the periods of 1980-1995 and 1995-2010. The estimation for the rate of change for the different covers was computed based on the formulae [8].

Carbon Quantification

A stratified sampling technique [12] based on vegetation cover classification was performed. Based on 2010 Landsat TM imagery, the vegetation covers obtained were closed forest, open forest, bushland and grassland. Each cover class was considered as a stratum. Settlements and other land uses class (e.g., bareland and cultivated land) were not included during inventory, as they had little vegetation.

Concentric plots [13] was adopted and used in this study. To ease the counting process, each sample plot was sub-divided into four sub-plots (concentric plots) of radius 2 m, 5 m, 10 m and 15 m. At the radius of 2 m, all trees with dbh <2 cm and >1 cm were recorded; at the radius of 5 m, all trees with dbh ≥ 2 cm and <10 cm were recorded; at the radius of 10m, all trees with dbh ≥ 10 and <20 cm were recorded; and at the radius 15 m, all trees with dbh ≥ 20 cm were recorded. Tree diameters were measured using a vernier caliper and tree heights were measured using a Suunto hypsometer. A botanist and local people were engaged for the identification of botanical/ scientific names and local names of trees respectively. The number of sample plots and the distance between plots were determined by the formula [14];

$$N = \frac{T_A x S_i}{P_s x100} \tag{3}$$

where N = number of sample plots, T_A = total area of the forest, P_s = plot size and S_i = sampling intensity, while the distance between plots was determined by the formula:

$$D = \sqrt{\left(\frac{Af x 10000}{N}\right)} \tag{4}$$

where D = inter plots distance (m), Af = area of the forest (ha) and N = number of plots.

The adopted sampling intensity was at least 0.1%. Therefore, for KFR, the total area of bushland, closed forest, grassland and open forest, based on the 2010 Landsat image classification was 4820.8 ha, making a total of 68 plots. While in PFR, the total area for bushland, closed forest, grassland and open forest was 2230.1 ha, making a total of 33 plots. Transects were created with concentric plots of radius 15 m (0.07 ha) systematically located at 842 m and 822 m intervals from each other in the North-South direction in KFR and PFR respectively (**Figure 2**), in each of the transects. According to [15], plots need to be allocated systematically so as to achieve a certain level of accuracy. During inventory, a GPS facilitated the orientation of the plots. In each plot, dbh, height and names (local and botanical names) of each tree were documented. Dbh was measured at 1.3 m above the ground level [13].

Tree above ground biomass (AGB) was computed as a product of total tree volume and wood basic density. The average wood density of 0.58 $g{\cdot}cm^{-3}$ for natural forest was used [16]. The volumes of trees were estimated using the formula [15]:

$$V_i = 0.5gh_i \tag{5}$$

Where V_i = volume of the ith tree (m^3)

g = the tree basal area (m^2)

0.5 = tree form factor.

The value recommended in natural forests of Tanzania without distinction of the vegetation type involved [17].

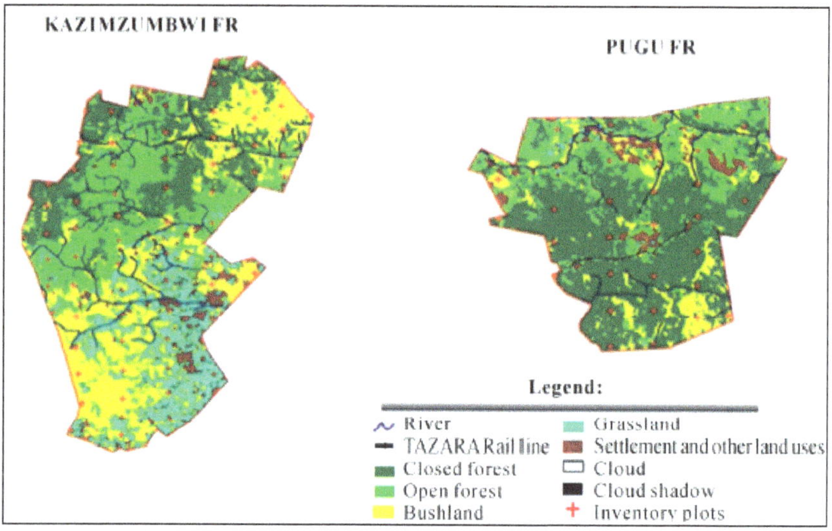

Figure 2: The map indicating KFR and PFR inventory plots.

The value of tree biomass was converted to carbon using a biomass-carbon ratio of 0.49 [12,18]. The carbon density for the whole forest was obtained by averaging carbon density from each individual forest stratum. Carbon stock was obtained by summing the products of the stratum's carbon density and their corresponding cover area. The carbon stocks for 1980 and 1995 were obtained by assuming that the carbon densities of each individual cover class did not change [19].

Carbon Mapping

According to [20], ordinary kriging using an exponential semivariogram model was considered to have the best performance for AGB estimation and for examining its spatial heterogeneity. Therefore, in this study, mapping of above ground carbon from trees was derived using ordinary kriging using an exponential semivariogram model.

RESULTS AND DISCUSSION

Accuracy Assessment for PKFR

The results from the classification accuracy assessment revealed that the overall accuracy of the classification for PFR was 84.85% and that of KFR was 82.35%. According to [21], the overall accuracy is acceptable if it is greater than 80%.

KFR Land Cover Maps and Their Changes

The land cover maps for the period 1980, 1995 and 2010 are presented in Figure 3, while Table 2 presents the cover areas for the respective periods and the area changes between the periods of 1980-1995 and 1995-2010. Generally, the maps indicate that there is significant variation between the periods under consideration.

Results (Table 2) indicate that in 1980, closed forest occupied 4050.9 ha (75.7%), bushland 728.8 ha (13.6%), grassland 269.4 (5%), open forest 260.2 (4.9%), settlement and other land uses 4.7 ha (0.1%). In 1995, closed forest occupied 3415.4 ha (63.8%) followed by grassland 741.2 ha (13.9%), open forest 654.3 ha (12.2%), bushland 449.1 ha (8.4%), and settlement and other land uses 89.7 ha (1.7%). Likewise, in 2010, closed forest occupied 1740.55 (32.5%), bushland 1131.7 ha (21.2%), open forest 1032.3 ha (19.3%), grassland 916.5 ha (17.1%) and settlement and other land uses 528.9 ha (9.9%).

During the period of 1980-1995, the result (Table 2) revealed that closed forests decreased by 635.5 ha (−11.9%) and 1674.9 ha (−31.3%) for the period of 1995-2010. Similarly, in the period of 1980-1995, settlement and other land uses increased by 85 ha (1.6%) and 439.2 ha (8.2%) between 1995-2010.

PFR Land Cover Maps and Their Changes

Figure 4 presents the land cover maps of PFR for the years 1980, 1995 and 2010 respectively, while Table 3 presents the cover areas for the respective years and the area changes between 1980 and 1995, and between 1995 and 2010. During the year 1980, it was revealed that closed forest occupied 2106.6 ha (87.2%), open forest 110.4 ha (4.6%), bushland 81.1 ha (3.4%), grassland 21.5 ha (0.9%), and settlement and other land uses 10 ha (0.4%) (Table 3). In 1995, closed forest occupied 1997.4 ha (82.7%), open forest 299.1 ha (12.4%), bushland 15.5 ha (0.6%, Grassland 18.3 ha (0.8%), and settlement and other land uses 85 ha (3.5%). Similarly, in 2010, closed forest occupied 1386.3 ha (57.4%), open forest 490.2 ha (20.3%), bushland 311.9 ha (12.9%), grassland 41.7 ha (1.7%), and settlement and other land uses 185 ha (7.7%).

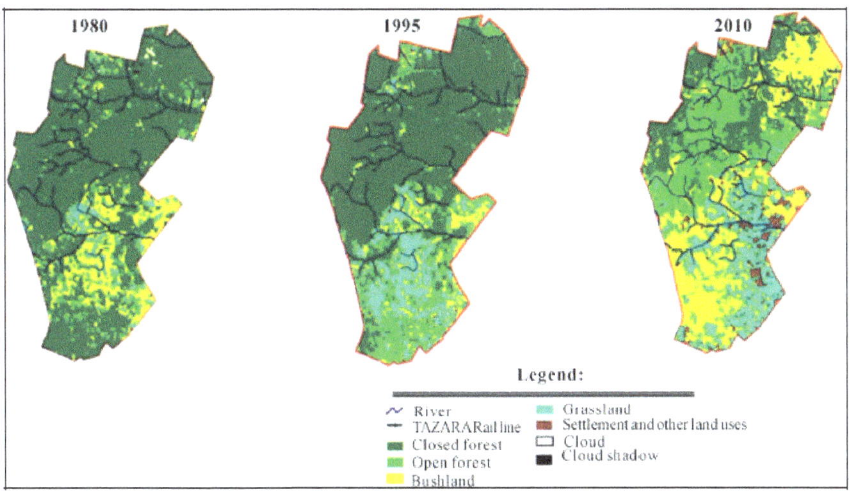

Figure 3: KFR land cover map for the years 1980, 1995 and 2010.

Table 2: Cover area, change area and annual rate of change between 1980 and 2010 for KFR

	1980		1995		2010		Change area (ha)	Change area (ha)	% Change	% Change	Annual rate of change (ha/year)	Annual rate of change (ha/year)
Cover class	Cover area (ha)	% Cover coverage	Cover area (ha)	% Cover coverage	Cover area (ha)	% Cover coverage	(1980-1995)	(1995-2010)	(1980-1995)	(1995-2010)	(1980-1995)	(1995-2010)
CF	4051	75.7	3415.4	63.8	1740.5	32.5	−635.5	−1674.9	−11.9	−31.3	−42.4	−111.7
OF	260.2	4.9	654.3	12.2	1032.3	19.3	394.1	378	7.4	7.1	26.3	25.2
BU	728.8	13.6	449.1	8.4	1131.7	21.2	−279.7	682.5	−5.2	12.8	−18.6	45.5
GL	269.4	5	741.2	13.9	916.5	17.1	471.8	175.2	8.8	3.3	31.5	11.7
SM	4.7	0.1	89.7	1.7	528.9	9.9	85	439.2	1.6	8.2	5.7	29.3
CL	26.9	0.5	0	0	0	0	−26.9	0	−0.5	0	−1.8	0
CS	8.8	0.2	0	0	0	0	−8.8	0	−0.2	0	−0.6	0
Total	5350	100	5349.7	100	5349.7	100						

Note: CF = Closed Forest; OF = Open Forest; BU = Bushland; GL = Grassland; SM = Settlement and other land uses; CL = Cloud; CS = Cloud Cover.

During the period of 1980-1995 (Table 3), closed forest decreased by 109.2 ha (−4.5%) and 611.1 ha (−25.3%) for the period 1995-2010. Similarly, for the period of 1980-1995, settlement and other land uses increased by 75 ha (3.1%) and 100.2 ha (4.1%) between 1995 and 2010. The findings also show that between 1980 and 2010, closed forest decreased, while bushland, grassland, settlement and other land uses, and open forest increased.

Generally, during the periods of 1980-1995 and 1995- 2010, the forest reserves have shown a significant degradation. Examples of the activities that contributed to the degradation of the forest reserves include charcoal burning and logging [22], encroachment for agriculture and pole cutting [5] and the expansion of farms within the forest reserves [23]. Also, [24] noted that the land use and cover changes in PKFR has been driven by the impacts of climate change and variability, which have intensified the resource extraction from the forest reserves.

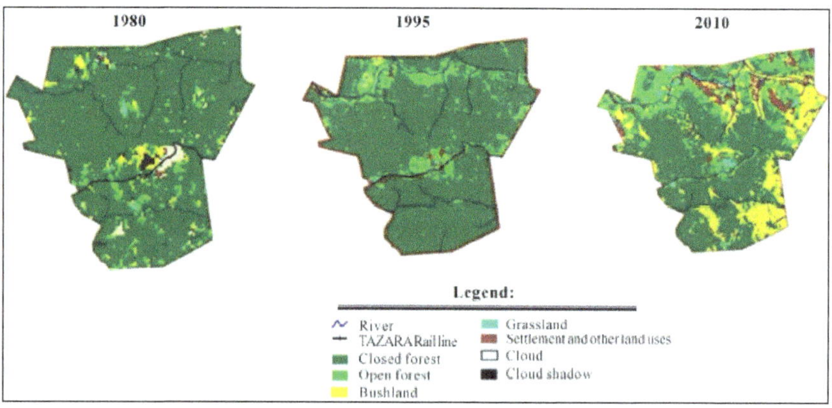

Figure 4: PFR land cover map for the year 1980, 1995 and 2010.

Table 3: Cover area, change area and annual rate of change between 1980 and 2010 for PFR

	1980	1980	1995	1995	2010	2010	Change area (ha)	Change area (ha)	% Change	% Change	Annual rate of change (ha/year)	Annual rate of change (ha/year)
Cover class	Cover area (ha)	% Cover coverage	Cover area (ha)	% Cover coverage	Cover area (ha)	% Cover coverage	(1980-1995)	(1995-2010)	(1980-1995)	(1995-2010)	(1980-1995)	(1995-2010)
CF	2106.6	87.2	1997.4	82.7	1386.3	57.4	−109.2	−611.1	−4.5	−25.3	−7.3	−40.7
OF	110.4	4.6	299.1	12.4	490.2	20.3	188.7	191.1	7.8	7.9	12.6	12.7
BU	81.1	3.4	15.5	0.6	311.9	12.9	−65.6	296.5	−2.7	12.3	−4.4	19.8
GL	21.5	0.9	18.3	0.8	41.7	1.7	−3.2	23.4	−0.1	1	−0.2	1.6
SM	10	0.4	85	3.5	185.2	7.7	75	100.2	3.1	4.1	5	6.7
CD	56.2	2.3	0	0	0	0	−56.2	0	−2.3	0	−3.7	0
CS	29.4	1.2	0	0	0	0	−29.4	0	−0.2	0	−2	0
Total	**2415.3**	**100**	**2415.3**	**100**	**2415.3**	**100**						

Note: CF = Closed Forest; OF = Open Forest; BU = Bushland; GL = Grassland; SM = Settlement and other land uses; CL = Cloud; CS = Cloud Cover.

Carbon Stock in PKFR

The results of the carbon densities revealed that closed forest had the highest mean carbon density compared to other cover classes. In KFR, the carbon density for closed forest was 2.29 tC/ha and 10.12 tC/ha for PFR. The results for other cover classes are summarized in Table 4. In KFR, the slightly higher mean carbon density in grassland is the result of several big trees in the plots, including *Mangifera indica* (Mwembe), *Vitex donian*a (Mfulu) and *Cyphostemma sp* (Mtamba). The study revealed that the overall mean carbon density for PFR and KFR were 5.72 tC/ha and 0.98 tC/ha respectively. Similarly, the 1980 carbon stock was found to be higher compared to 1995 and 2010, for both forests.

Table 4: Carbon density for different cover classes in PKFR

Forest reserve	Carbon density (tC/ha)			
	BU	OF	CF	GL
PFR	1.82	0.24	10.12	0
KFR	0.21	0.36	2.29	1.52

Note: BU = Bushland; OF = Open Forest; CF = Closed Forest; GL = Grass-land.

The study revealed that 1980 carbon stock in PFR was 21 496.5 tC/ha, while in KFR it reached 12 768.4 tC/ha (Table 5). According to [25], the total carbon stock in living forest biomass in Tanzania was estimated at 2019 million tons.

Carbon Map

The carbon maps for KFR and PFR are presented in Figures 5 and 6.

Table 5: Carbon stock for PKFR for the year 1980, 1995 and 2010

Year	Carbon stock (tC/ha)	
	PFR	KFR
1980	21,496.5	12,768.4
1995	20,314.3	11,668.5
2010	14,730.4	7206.5

Generally, the study revealed that the mean carbon densities for PFR and KFR were 5.72 and 0.98 tC/ha, respectively. According to [26], the average carbon density of the remaining coastal forests in Tanzania is estimated to be 64 tC/ha. The contribution of PKFR to the overall carbon density of coastal forests seems to be lower. Likewise, the estimated carbon densities in the forest reserves were low compared to 58-94 tC/ha for the Arabuko Sokoke coastal forest of Kenya [27] and 16.5 tC/ha for the miombo woodlands of the Kitulangalo forest reserve [28].

CONCLUSIONS

The study has revealed that the forest reserves conditions have changed extensively between the years of 1980 and 2010.

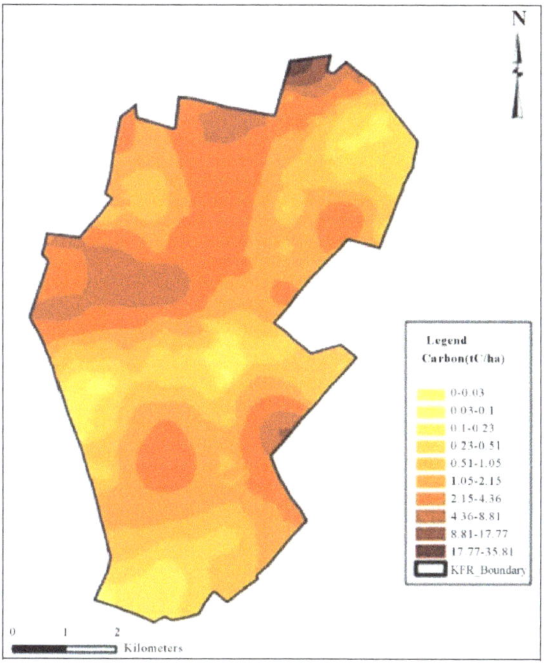

Figure 5: Carbon density map for KFR.

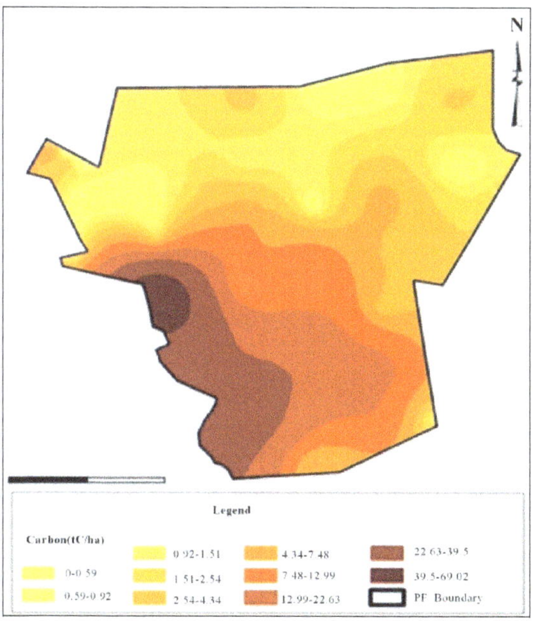

Figure 6: Carbon density map for PFR.

Closed forest has decreased while bushland, grassland, open forest and settlement and other land uses have increased. Between 1980 and 1995, and between 1995 and 2010, closed forest decreased by 11.9% and 31.3% respectively in KFR. During the same period, closed forest decreased by 4.5% and 25.3% respectively in PFR. The mean carbon densities were low, and estimated to 0.98 tC/ha and 5.72 tC/ha for KFR and PFR respectively. It was also revealed that forest reserves have shown a potential in carbon storage, estimated at 7206.46 tC and 14 730.41 tC for KFR and PFR respectively.

The low carbon densities are thought to decline even more in areas under closed forests in the two forest reserves. The study recommends concerted efforts to enhance proper management of the forests and increased environmental education for communities so that the two forest reserves may contribute to REDD initiatives.

ACKNOWLEDGEMENTS

This research was supported by a NORAD funded research programme—Climate Change Impacts, Adaptation and Mitigation (CCIAM) of the Sokoine University of Agriculture through Analysis of the Impacts of Urban Land Use and Climate Change on Coastal Forest Eco- system and Management research project.

REFERENCES

1. T. Herzog, "World Greenhouse Gas Emissions in 2005," WRI Working Paper, 2009. http://pdf.wri.org/working_papers/world_greenhouse_gas_emissions_2005.pdf
2. M. Grieg-Gran, "Is Tackling Deforestation a Cost-Effective Mitigation Approach?" 2006. http://pubs.iied.org/
3. R. A. Houghton, "Revised Estimates of the Annual Net Flux of Carbon to the Atmosphere from Changes in Land Use and Land Management," *Tellus*, Vol. 55B, No. 2, 2003, pp. 378-390.
4. N. Myers, R. A. Mittemeier, C. G. Mittemeier, J. Kent and A. B. Gustavo, "Biodiversity Hotspot for Conservation Priorities," *Nature*, Vol. 403, No. 6722, 2000, pp. 853-858. doi:10.1038/35002501
5. M. Hogqvist, S. Nummelin, M. Stahl and B. Miljoventenskap, "The View of Nature of Villagers in Kisarawe District," The Department of Nature and Environment, Karlstad University, Karlstad, 1998.
6. United Republic of Tanzania (URT), "National Bureau of Statistics," Government Printers, Dar es Salaam, 2002.
7. B. P. Mbilinyi, "Assessment of the Land Degradation and Its Consequences: Use of Remote Sensing and Geographical Information System Techniques. A Case Study in the Ismani Division, Iringa Region, Tanzania," Ph.D. Thesis, Berlin Technical University, Berlin, 2000.
8. J. J. Kashaigili and A. M. Majaliwa, "Integrated Assessment of Land Use and Cover Changes in the Malagarasi River Catchment in Tanzania," *Physics and Chemistry of the Earth*, Vol. 35, No. 13-14, 2010, pp. 730-741. doi:10.1016/j.pce.2010.07.030

9. Atesoglu and M. Tunay, "Spatial and Temporal Analysis of Forest Cover Changes in Bartin Region of North-West Turkey," *African Journal of Biotechnology*, Vol. 9, No. 35, 2010, pp. 5676-5685.

10. R. G. Congalton, "A Review of Assessing the Accuracy of Classifications of Remotely Sensed Data," *Remote Sensing Environment*, Vol. 37, No. 1, 1991, pp. 35-46. doi:10.1016/0034-4257(91)90048-B

11. M. Reusing, "Change Detection of Natural High Forests in Ethiopia Using Remote Sensing and GIS Techniques," *International Archives of Photogrammetry and Remote Sensing*, Vol. XXXIII, Part B7, 2000, pp. 1253-1258. http://www.isprs.org/proceedings/xxxiii/congress/part7/1253_XXXIII-part7.pdf

12. K. G. MackDicken, "A Guide to Monitoring Carbon Storage in Forest and Agroforestry Projects," Winrock Internation Institute for Agricultural Development, Little Rock, 1997.

13. National Forest Resources Monitoring and Assessment of Tanzania-NAFORMA, "Field Manual Biophysical Survey," Forest and Beekeeping Division, Dar es Salaam, Tanzania, 2010.

14. P. K. T. Munishi, "Principal of Forest Ecology. A Compendium-SUA," 2005.

15. E. Zahabu, "Sinks and Sources: A Strategy to Involve Forest Communities in Tanzania in Global Climate Policy," Ph.D. Thesis, University of Twente, Enschede, 2008.

16. P. K. T. Munishi and T. Shear, "Carbon Storage of Two Afromontane Rain Forests in the Eastern Arc Mountains of Tanzania," *Journal of Tropical Forest*, Vol. 16, No. 1, 2004, pp. 78-93.

17. E. F. Haule and F. C. Munyuku, "National Forest Inventory in Tanzania," In: R. E. Malimbwi and E. J. Luoga, Eds., *Proceedings of the workshop on Information Acquisition for Sustainable Natural Forest Resources of Eastern, Central and Southern Africa*, Faculty of Forestry, SUA, Morogoro, 1994, pp. 99-113.

18. S. Brown, "Measuring, Monitoring and Verification of Carbon Benefits for Forest Based Projects," In: I. R. Swingland, Ed., *Capturing Carbon and Conserving Bio-diversity: The Market Approach*, Earthscan Publications Ltd., London, 2003, pp. 118-133.

19. H. K. Gibbs, S. Brown, J. O. Niles and J. A. Foley, "Monitoring and Estimating Tropical Forest Carbon Stocks: Making REDD a Reality," *Environmental Research Letter*, Vol. 2, No. 4, 2007, p. 13.

20. D. Huaqiang, Z. Guomo, F. Wenyi, G. Hongli, X. Xiao- jun, S. Yongjun and F. Weiliang, "Spatial Heterogeneity and Carbon Contribution of Aboveground Biomass of Moso Bamboo by Using Geostatistical Theory," *Plant Ecology*, Vol. 207, No. 1, 2010, pp. 131-139. sdoi:10.1007/s11258-009-9659-3

21. S. O. Turan, A. I. Kodiogullari and A. Gunlu, "Spatial and Temporal Dynamics of Land Use Pattern Response to Urbanization in Kastamonu," *African Journal of Bio-technology*, Vol. 9, No. 5, 2010, pp. 640-647.

22. Ahrends, "Patterns of Degradation in Lowland Coastal Forests in Coast Region, Tanzania," MSc Thesis, Greifswald University, Greifswald, 2005.
23. O. E. Malugu, "Resources-Use Conflicts and Management Challenges for Pugu and Kazimzumbwi Forest Reserves in Kisarawe and Ilala Districts, Tanzania," *Discovery and Innovation*, Vol. 19, No. 1-2, 2007, pp. 149- 174. doi:10.4314/dai.v19i1-2.15781
24. M. V. Mdemu, J. J. Kashaigili, J. Lupala, P. Levira, E. Liwenga, A. Nduganda and F. Mwakapuja, "Dyanamics of Land Use and Land Cover Changes in the Pugu and Ka- zimzumbwi Forest Reserve," IULUCC Project, Ardhi University, Dar Es Salaam, 2012.
25. FAO, "State of the World's Forests," Rome, 2011.
26. F. L. Godoy, K. Tabor, N. D. Burgess, B. P. Mbilinyi, J. J. Kashaigili and M. K. Steininger, "Deforestation and CO2 Emissions in Coastal Tanzania from 1990 to 2007," *Environmental Conservation*, Vol. 39, No. 1, 2011, pp. 62-71. doi:10.1017/S037689291100035X
27. G. Julia, "Carbon Storage and Emissions Offset Potential in an African Dry Forest, the Arabuko-Sokoke Forest, Kenya," *Environmental Monitoring and Assessment*, Vol. 142, No. 1-3, 2008, pp. 85-95.
28. R. E. Malimbwi, B. Solberg and E. Luoga, "Estimation of Biomass and Volume in Miombo Woodland at Kitulangalo Forest Reserve," *Tanzania Journal of Tropical Forest Science*, Vol. 7, No. 2, 1994, pp. 230-242.

8

CHAPTER

EVALUATING BIOMASS OF JUNIPER TREES (JUNIPERUS PINCHOTII) FROM IMAGERY DERIVED CANOPY AREA USING THE SUPPORT VECTOR MACHINE CLASSIFIER

Mustafa Mirik, Sriroop Chaudhuri, Brady Surber, Srinivasulu Ale and Robert James Ansley

Keywords: Remote Sensing, Geospatial Imagery, Biomass, Juniperus pinchoti, Rangeland, Bioenergy, Terrestrial Carbon Budget

INTRODUCTION

Aboveground vegetation biomass (biomass hereafter) distribution, structure, and amount in rangeland regions are important in estimating terrestrial carbon and energy fluxes between the atmosphere and biosphere, and thus are major considerations in ecosystem processes or climate change research [1-4]. Vegetation on rangelands, characterized by woodlands, shrublands, and grasslands, provides essential ecosystem services, including forage for domestic and native herbivores, but is highly variable in spatial distribution and temporal growth patterns [5-9]. A world-wide phenomenon on many rangelands has been the expansion of woody plants into grassland and savanna areas in the last 150 years [3,4,8,10]. Increases in woody-plant abundance make up a significant but highly uncertain fraction of the carbon budget world-wide [3,6]. Projecting future states of rangeland ecosystems in response to woody invasions and changes in climate, edaphic, and disturbance regimes are strongly dependent on current assessment of woody biomass and distribution, and the contribution of woody plants to total terrestrial biomass amount [6].

Traditional field based biomass estimations of woody biomass through allometric functions are accurate, but are costly, tedious, time consuming, and usually limited

Citation: M. Mirik, S. Chaudhuri, B. Surber, S. Ale and R. Ansley, "Evaluating Biomass of Juniper Trees (Juniperus pinchotii) from Imagery-Derived Canopy Area Using the Support Vector Machine Classifier," Advances in Remote Sensing, Vol. 2 No. 2, 2013, pp. 181-192, doi: 10.4236/ars.2013.22021.

to small land areas. Remote sensing based biomass prediction is the most promising method at regional and global scales [2,11-18]. Over the last decade, a large number of studies have estimated biomass using two dimensional (2D) geospatial images, synthetic aperture radar (SAR) and LIDAR [2,11,13,16,19]. Numerous studies have documented that remote sensing techniques are highly efficient in assessing woody plant biomass at multiple spatial and temporal scales [11,12,20]. Although small footprint radar backscatter and LIDAR data may be more accurate than 2D geospatial imagery for characterizing woody plant attributes, this type of data is expensive, and is limited by its narrow swath coverage on each overpass due to high spatial resolution [4,15,21-23]. Two dimensional high resolution geospatial imageries, such as TM, IKONOS, QuickBird, GeoEye and WorldView, are much more cost-effective and are now available for almost anywhere in the world [24]. Though such data has been widely used to quantify forest biomass, it has been rarely utilized for rangeland woody biomass assessment [24,25]. Research on high-resolution 2D geospatial images from satellites has included the evaluation of algorithms to determine crown diameter, crown area, and other attributes in a wide range of forested areas [24,25]. For this reason, remote sensing studies quantifying rangeland biomass contribution to the total carbon budget are strongly needed.

Redberry juniper (*Juniperus pinchotii*; juniper hereafter) occupies more than 4.7 million ha of rangeland in the western half of Texas alone and extents to the neighbouring states in the Southern Great Plains of the US [10]. Juniper is a native troublesome species, aggressively encroaching into rangeland due to reduced intensity and frequency of fires and livestock overgrazing. High densities of this species, and a closely related species in central TX, Ashe juniper (*Juniperus ashei*), can significantly decrease species diversity and grass production by competing for natural resources and can greatly trigger soil erosion [26].

The United States Department of Agriculture-National Agricultural Imagery Program (USDA-NAIP) imagery archive has entered public domain, providing exciting opportunities to advance rangeland research. We propose that periodically acquired NAIP images are adequate for quantifying biomass of individual juniper plants because of the 1-m spatial resolution that is available. Recently, allometric models between field measured redberry juniper above ground mass (AGM) and physiognomic characteristics such as canopy area, diameter, and height were developed [10]. In addition, the potential of high-resolution QuickBird images to estimate *Juniperus* spp. tree volume and biomass was examined to determine if this type of optical imagery can be used for rapid and reliable measurements of tree volume and biomass [27,28]. The objectives of the present study were to 1) identify and delineate juniper from surrounding live vegetation using a support vector machine (SVM) classifier; 2) investigate the relationships between imagery derived and field measured canopy area and diameter; and 3) develop regression models relating imagery derived and field measured single tree attributes for an estimation of total, wood, and leaf-twig biomass. Since the remote sensing biomass of juniper trees has not been well documented, we initiated this study to fill an important gap in the application of remotely assessed canopy area using the SVM method for discriminating juniper plants from background vegetation and estimating biomass components.

MATERIALS AND METHODS

Study Area

Data was collected from two sites near the town of Crowell, approximately 132-km west of Wichita Falls and 241-km east of Lubbock, in north central Texas (Figure 1). The sites were located within 10 km of each other on similar soil types: Site 1 (Whatley Ranch), 34°4'N, 99°44'W; elev. 436-m in Foard County; Site 2 (Copper Break State Park), 34°6'N, 99°45'W; elev. 443-m in Hardeman County. The total 30 years mean annual rainfall at Crowell is 616 mm [29]. The sites were dominated by juniper and honey mesquite (*Prosopis glandulosa* Torr.). Herbaceous understory consists of numerous cool season (C3) and warm season (C4) grass species dominated by perennial C4 grasses, tobosagrass (*Pleuraphis mutica*) and buffalograss (*Bouteloua dactyloides*). Other grasses include C3 mid-grass Texas wintergrass (*Nassella leucotricha*), C3 annual grass Japanese brome (*Bromus japonicus*) and other C4 mid-grasses, such as vine mesquite (*Panicum obtusum*), sideoats grama (*Bouteloua curtipendula*), and dropseeds (*Sporobolus* spp.) [30]. Soils are fine-silty, mixed, thermic Typic Calciustolls of the Quanah series and fine-silty, montmorillonitic, thermic Typic Haplusterts of the Hollister series [31].

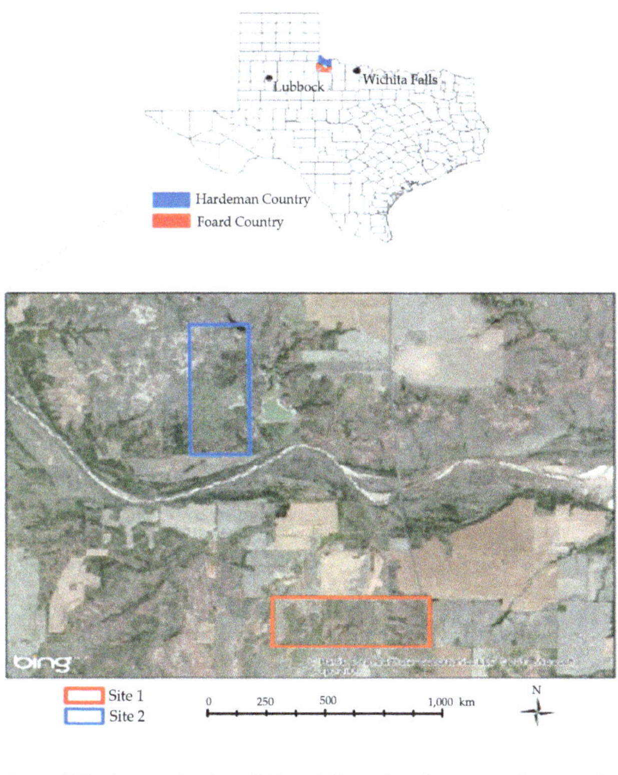

Figure 1: Location of Hardeman (top) and Foard Counties (bottom) where study sites were situated in north central Texas.

Field Sampling

A total of 54 juniper plants ranging in size from 1.53 to 73.90-m^2 canopy area were selected for the study. The trees were selected to represent the range of three diameters found at the sites, thus were not random because the objective of the study was to quantify the relationship between the tree attributes derived from the imagery and field measured biomass through regression analysis. To randomly sample the study sites and still have the same range of canopy diameter would have required many more plots than the logistics of the study could accommodate. Selection was limited to trees that appeared healthy and undamaged and were at least 1-m apart from other trees. Canopy diameters along 2 directions (longest axis and perpendicular to longest) were measured on each tree followed by harvesting to ground level during the dormant season (October-December 2010 and January-March 2011). Following harvest, leaf-twig (<3-cm diameter stems) and wood (>3-cm diameter) components were manually separated and weighed. Total wet weight of each component was obtained within a few hours of tree felling. Three randomly selected sub-samples of each component from each tree were weighed as wet weight, oven-dried at 60°C for 60 - 70 days, and weighed after weights stabilized. Percent moisture was determined [(wet weight-dry weight)/wet weight × 100] for components and as well as the whole tree and used to determine total oven dry biomass of each tree and individual biomass components.

Remote Sensing Imagery and Image Classification

One county-level colour infrared aerial image collected over Hardeman County, covering both sites, was obtained from the National Agricultural Imagery Program (NAIP) provided by the Natural Resources Conservation Service Geospatial Data Gateway at (http://datagateway.nrcs.usda.gov/). The NAIP imageconsisted of green, red, and near infrared bands with a spatial resolution of 1-m, taken on August 12, 2010 over the sites. The image was projected to the Universal Transverse Mercator North American Datum 1983 Zone 14 North by the provider was the sites were extracted using ArcGIS software suits (ESRI, Inc. Redlands, CA).

Image classifications were performed using the SVM method in Environment for Visualizing Images software (ENVI; Exelis Visual Information Solutions, Boulder, CO). The SVM is a supervised machine learning method that performs classification based on the statistical learning theory. The SVM classifies data by separating a hyperplane that provides the best separation between classes in a multidimensional feature space. This hyper-plane is the decision surface on which the optimal class separation takes place. The optimal hyperplane that which maximizes the distance between the hyperplane and the nearest positive and negative training example, known as the margin. From a given set of training samples, the optimization problem is solved to find the hyperplane that leads to a sparse solution. Although the SVM is a binary classifier in its simplest form, implementation of the SVM classifier in ENVI was extended to more than two classes by splitting the problem into a series of binary class separations.

In order to represent more complex shapes than linear hyperplanes, a variety of kernels, including the polynomial, the radial basis function, and the sigmoid, can be used for performing SVM classification in ENVI. The SVM was employed using the radial basis function kernel for performing the pairwise classification. There is also a penalty parameter can be introduced to the SVM classifier to allow for misclassification during the training process. The penalty parameter was set to its maximum value, whereas a classification probability threshold of zero was used in order to classify all pixels. Default settings of this classifier were used for image classifications. During the classification process, only the NIR band of the NAIP imagery was chosen using the spectral subset option in ENVI.

The 1-m image allowed for clear and visual identification of all endmembers based on the spectral contrast among the live vegetation (juniper, mesquite, herbaceous) and senescent or non-vegetative components (road, shadow, exposed soil, water, senescent herbaceous) (Figure 2(a)). Previous studies found that reflectance from Ashe juniper (*Juniperus ashei*), mesquite, water, exposed soil, and herbaceous plants were significantly different [32,33]. In addition, reflectance variation within deciduous crowns is larger than that within coniferous tree crowns because of the non-conical shape, large branches, and the shade area caused by the neighbouring branches [15]. Thus, juniper endmember was manually extracted from isolated trees on areas in the image [34]. Juniper endmember for each site comprised of 10 polygons, each having 25 pixels, from pure juniper canopy at locations well identified on the ground and on the image. The CEM was run for juniper endmember using a covariance matrix with a 75% rule threshold and square root stretching.

Determination of Canopy Area and Development of Regression Model for Biomass Estimation

Point coordinates of the harvested trees, with sub-meter accuracy at the centre of each tree following harvest, were taken with a Trimble GeoXH hand-held unit (Trimble Navigation Limited, Sunnyvale, CA) equipped with the ArcPad software (ESRI, Inc. Redlands, CA). The projection of the GPS unit was set to Universal Transverse Mercator North American Datum 1983 Zone 14 North. A unique identification number was assigned to each tree during the collection of point coordinates in the field. Spatial analysis performed for juniper canopy area determination is described below.

Following image classification using CEM, fractional abundance of juniper was converted to vector layer (polygon shapefile format) in ENVI and exported to ArcGIS for canopy area determination (Figure 2(b)). The point coordinates representing tree locations were overlaid onto the vector layer (Figure 2(c)) and the polygons representing harvested tree were selected with the "select by location" function, where the locations were tree point coordinates. A new polygon layer including only selected polygons was created (Figure 2(d)). This new polygon layer and point layer, both representing harvested trees, were "joined" in order to assigned unique identification number to each polygon. The area of each polygon was determined using the "calculate geometry" function and designated as canopy area. The attribute table of the circle layer was then exported to a spreadsheet in order to perform a regression analysis with field measured tree attributes.

The point coordinates representing tree locations were overlaid onto the image where one-to-one matching with the juniper trees was verified (Figure 2(c)). Circles representing juniper canopy area based on point locations were blindly digitized as a polygon layer by an untrained person in ArcGIS (Figure 2(c)). The areas and diameters of these circles were determined by the "calculate geometry" and "field calculator" functions, respectively.

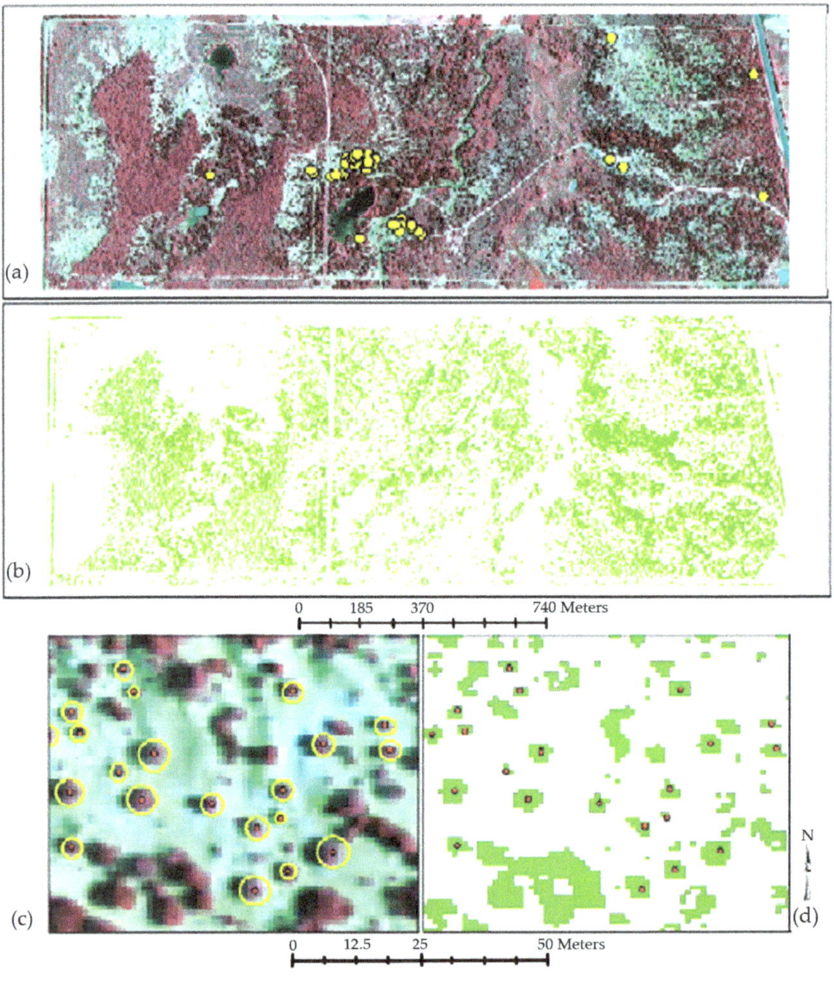

Figure 2: The false colour infrared national agricultural imagery program image with 1-m spatial resolution, harvested trees represented as yellow (a); classified image with juniper: green (b); some of the harvested trees illustrated with blindly drown circles in yellow and red dots (c); classified trees shown with the red dots (d) for site 1.

The attribute table of the circle layer was then exported to a spreadsheet in order to perform regression analysis with field measured tree attributes. Field based canopy area

(CA; m^2) determination was made using the ellipse equation: $\pi \cdot a \cdot b$, where a = radius of longest canopy axis and b = radius of perpendicular canopy axis adopted from [10].

Simple linear regression analyses were performed to relate image derived tree attributes with field measured tree variables using Statistical Analysis Systems 9.2 (SAS Institute Inc., Cary, NC) and regression plots were created using SigmaPlot 11 (Systat Software Inc., Chicago, IL). Image derived tree parameters were used as the independent variable and field measured attributes were set as the dependent variable. Model validation was performed by randomly apportioning observations into a 70/30 split, with the former fraction being used for model development and the latter used to predict tree attributes. Error estimation was performed by computing the root mean squared error (RMSE) between predicted and field measured values, as well as field measured and imagery derived attributes.

RESULTS

Data Summary

Descriptive summary of variables used in regression models were presented in Table 1. The means and ranges confirm that a broad range of values was assessed for tree attributes. Field measured juniper plant canopy diameters ranged from 1.40 to 9.70-m with a mean and standard error ($\bar{x} \pm \varepsilon$: 5.15 ± 0.31), whereas canopy areas varied from 1.72 to 73.90-m^2 (24.73 ± 2.72). Average canopy diameters and areas derived from the imagery using both manual extraction and classification methods were slightly higher than the field measurements (Table 1).

Table 1: Summary statistics for juniper tree attributes measured in the field and extracted from the imagery with 1-m spatial resolution (n = 54).

Tree attribute	Mean	SE	Min	Max
Fmca (m^2)	24.73	2.72	1.53	73.90
Fmcd (m^2)	5.15	0.31	1.40	9.70
Fmtb [kg]	187.03	25.29	3.53	688.35
Fmwb [kg]	104.78	16.41	0.05	460.37
Fmltb [kg]	82.25	9.47	3.44	227.99
Meca (m^2)	28.21	3.27	1.72	101.92
Macd (m^2)	5.48	0.33	1.48	11.39
Cca (m^2)	27.46	3.55	0.00	105.00

Fmca: field measured canopy area; Fmcd: field measured canopy diameter; Fmtb: field-measured total biomass; Fmwb: field measured wood biomass; Fmltb: field measured leaf-twig biomass; Meca: manually extracted canopy area; Mecd: manually extracted canopy diameter; Cca: classified canopy area; SE: Standard Error; Min: Minimum; Max: Maximum.

Canopy diameter and area for the manual extraction method ranged from 1.48 to 11.39 m (5.48 ± 0.33) and from 1.72 to 101.92-m^2 (22.84 ± 2.70), respectively. Range of canopy areas derived from classified imagery were between 0 and 105-m^2 (27.46 ± 3.55), respectively.

A juniper plant with 1.75 m diameter, and 2.3 m^2 canopy area was, however, not detected with the algorithm. Biomass varied from 3.53 to 688.85 kg (187.03 ± 24.83). Biomass was partitioned into wood biomass only with minimum and maximum weights 0.05 and 460.37 kg (104.78 ± 16.41), and leaf-twig biomass varying from 3.44 to 227.99 kg (82.25 ± 9.47).

Biomass Estimation

Simple linear regression analysis revealed the presence of robust, tight, and significant relationships between the variables of interest (Figures 3-6, Table 2). Predictors (classified and manually extracted canopy area and diameter) accounted for 91% to 96% of the variation in field measured canopy area and diameter with associated RMS errors between 4.274 and 0.489, respectively, (Figure 3). Manually extracted canopy area produced a slightly stronger relationship with total and wood biomass compared to canopy diameter (Table 2). Classified canopy area had equal predictive power for total and leaf-twig biomass (Figures 4(a)-(c)), while the predictive power was weaker for wood biomass (Figure 4(b)). The strength of the relationships between biomass components and imagery derived canopy area or canopy diameter generally declined (r^2 = 0.88 and RMSE between 5.754 and 5.106) for leaf-twig biomass (Figure 6(d); Table 2), except for the classified canopy area (r^2 = 0.93; RMSE = 7.751; Figure 4(c)).

The relationships between remotely determined and field measured tree attributes using the validation dataset confirmed robust and significant associations (Figures 5 and 6, Table 2). Linear regression between canopy diameter using the manual extraction method and the field measured canopy diameter produced a stronger relationship (r^2 = 0.95, RMSE = 0.463) (Figure 5(c)) compared to canopy area calculated using both manually extracted (r^2 = 0.94, RMSE = 4.287; Figure 5(a)) and classified data (r^2 = 0.94, RMSE = 4.199; Figure 5(b)), respectively. Total, wood, and leaf-twig biomass determined using classified canopy area accounted for 98% (RMSE = 16.644), 96% (RMSE = 15.717), and 88% (RMSE = 12.495) of the variations in field measured counterparts (Figures 6(a)- (c)), respectively. Biomass components estimated using manually extracted canopy area had r^2 values of 0.96, 0.96, and 0.84 for field measured total, wood, and leaf-twig biomass (Table 2), respectively, whereas the relationships between field measured total, wood, and leaf-twig biomass and biomass of these components determined using manually extracted canopy diameter were 93%, 90%, and 89% (Table 2), respectively.

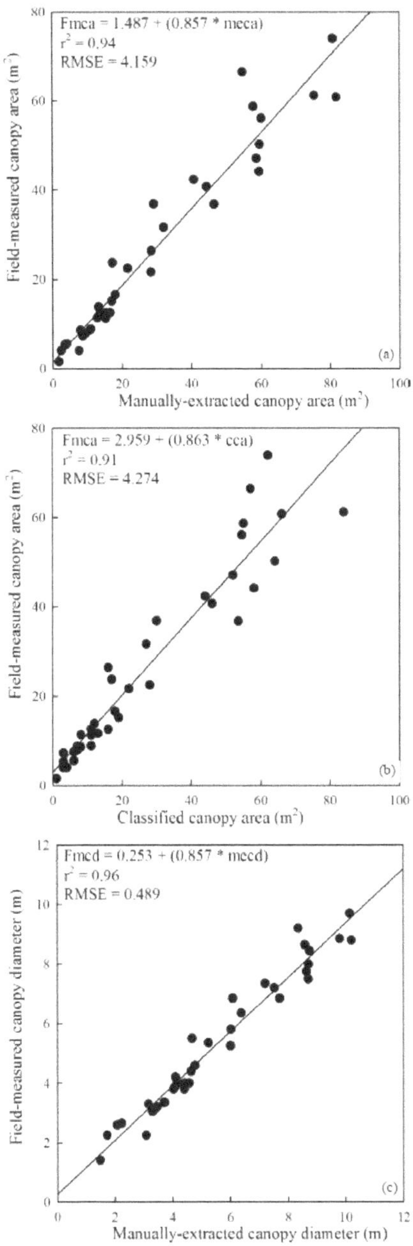

Figure 3: The relationship between manually extracted and field measured canopy area (a), classified and field measured canopy area (b), and manually extracted and field measured canopy diameter (c); n = 38, $P < 0.001$. Fmca: field measured canopy area; meca: manually extracted canopy area; cca: classified canopy area; fmcd: field measured canopy diameter; mecd: manually extracted canopy diameter.

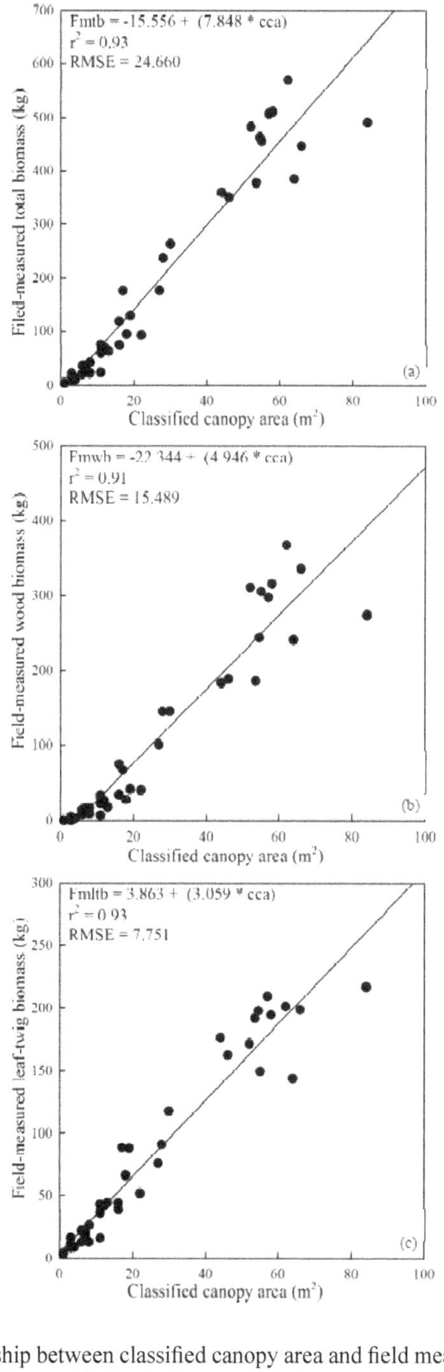

Figure 4: The relationship between classified canopy area and field measured total (a), wood (b), and leaf-twig biomass (c); n = 38, $P < 0.001$. Fmtb: field measured total biomass; cca: classified canopy area; fmwb: field measured wood biomass; fmltb: field measured leaf-twig biomass.

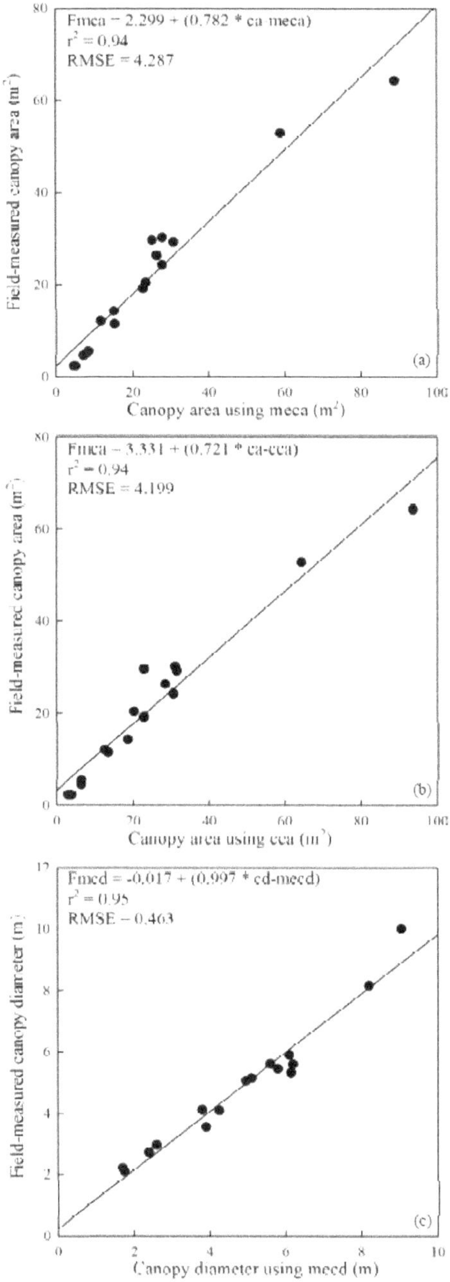

Figure 5: The relationship between canopy area using manually extracted canopy area (meca) and field measured canopy area (fmca) (a); canopy area using classified canopy area (cca) and fmca (b); and canopy diameter using manually extracted canopy diameter (mecd) and field measured canopy diameter (fmcd) (c); n = 16, P < 0.001.

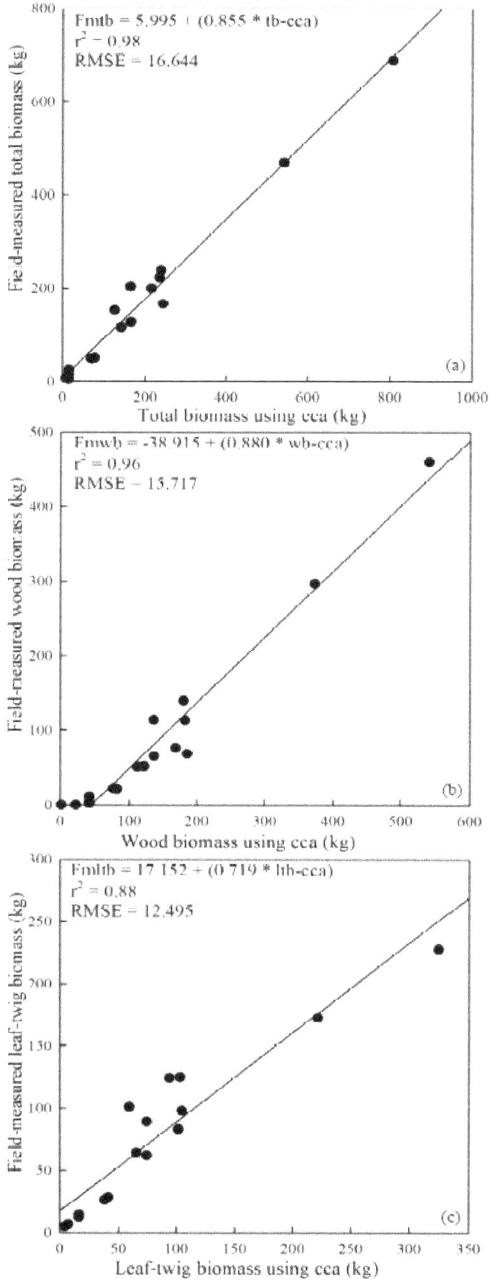

Figure 6: The relationship between total (a), wood (b), and leaf-twig biomass (c) determined from classified canopy area (cca) and field measurements; n = 16, P < 0.001. Fmtb: field measured total biomass; fmwb: field measured wood biomass; fmltb: field measured leaf-twig biomass; tb-, wb-, ltb-cca: total, wood, and leaf-twig biomass using classified canopy area.

Table 2: The simple linear regression models with the associated statistics for manually extracted canopy area and diameter versus predicted and field measured total, wood, and leaf-twig biomass.

Biomass model	r^2	RMSE
$\text{Fmtb}^{+} = -23.617 + (7.604^{*}\text{meca})$	0.92	24.87
$\text{Fmwb}^{+} = -30.581 + (4.903^{*}\text{meca})$	0.93	11.58
$\text{Fmltb}^{+} = 2.310 + (2.909^{*}\text{meca})$	0.88	5.75
$\text{Fmtb}^{+} = -202.34 + (71.921^{*}\text{mecd})$	0.90	29.56
$\text{Fmtb}^{+} = -142.143 + (45.704^{*}\text{mecd})$	0.89	20.16
$\text{Fmltb}^{+} = -67.776 + (27.822^{*}\text{mecd})$	0.88	5.10
$\text{Fmtb}^{*} = 8.674 + (1.457^{*}\text{fmtb-meca})$	0.96	16.96
$\text{Fmwb}^{*} = -11.972 + (0.992^{*}\text{fmwb-meca})$	0.96	5.23
$\text{Fmltb}^{*} = 12.208 + (0.800^{*}\text{fmltb-meca})$	0.84	5.65
$\text{Fmtb}^{*} = -42.420 + (1.085^{*}\text{fmtb-mecd})$	0.93	28.21
$\text{Fmwb}^{*} = -44.061 + (1.184^{*}\text{fmwb-mecd})$	0.90	18.48
$\text{Fmltb}^{*} = -116.266 + (0.889^{*}\text{fmltb-mecd})$	0.89	1.54

+ and *: model development ($n = 38$, $P < 0.001$) and validation datasets ($n = 16$, $P < 0.001$), respectively; fmtb: field measured total biomass; meca: manually extracted canopy area; fmwb: field measured wood biomass; fmltb: field measured leaf-twig biomass; mecd: manually extracted canopy diameter; ptb-meca, pwb-meca, and pltb-meca: predicted total, wood, and leaf-twig biomass using manually extracted canopy area, respectively; ptb-mecd, pwb-mecd, pltb-mecd: predicted total, wood, and leaf-twig biomass using manually extracted canopy diameter.

DISCUSSION

We evaluated and reported the usefulness of 2D geospatial aerial imagery for estimating juniper tree biomass at two rangeland sites, typical for the southern Great Plains of the US, using both automated and manual canopy area determination methods. Although accuracy assessment for classification was not conducted in this research, earlier studies found a high classification accuracy for juniper species [32,33,35-38]. Overall accuracies between 80 and 92% were found for classifying Ashe juniper in central Texas using QuickBird imagery and aerial photography [32]. An airborne hyperspectral image was evaluated in order to compare classification methods for mapping Ashe juniper in central Texas [33]. The authors found that overall mapping accuracies varied from

84% to 100%. Overall classification accuracies between 80% and 92% were achieved for both Rocky Mountain juniper (*Juniperus scopulorum*) and Utah juniper (*Juniperus osteosperma*) using Landsat 5 TM imagery [38]. A mapping accuracy of 92% for western juniper (*Juniperus occdentalis* subsp. *Occidentalis*) using the NAIP imagery was reported [35].

A single juniper plant with a 2.3-m^2 canopy area (1.75-m canopy diameter) was not detected with the algorithm used to classify the imagery in this study. Visual inspection of ground-level images of this plant taken prior to harvest indicated that it was growing on 100% exposed soil without any herbaceous background cover. This suggested that 100% exposed soil background may suppress detection of smaller plants. Accurate monitoring of semi-arid vegetation using remote sensing can be hindered by the effects of bare soil [39]. Reasons for this may relate to the greater level of "brightness" associated with bare soil in arid and semi-arid landscapes that saturates that portion of the image and eliminates detection of small dark objects such as a small tree canopy.

The approach developed and implemented in the current study allows accurate estimation of individual juniper tree biomass using simple linear regression for field measured biomass and automated or manual methods. The relationships observed between field measured and image derived juniper plant variables were robust and highly significant. This indicates that remotely sensed imagery can be very useful for estimating biomass of individual juniper trees on rangeland ecosystems for both trained and untrained personnel. For example, an automated canopy area determination method over large land areas can be used by trained personnel in image processing software, while the manual method can be preferable by untrained rangeland managers. Remote sensing has not been used extensively for quantifying the amount and extent of woody biomass on rangelands. The bulk of the existing studies have relied on other mass quantification techniques, which usually are time consuming measurements using field transects [9,28]. The strength of the relationships found in this study implies that these attributes can be readily obtainable from the image with a sufficient spatial resolution without a need for field measurements, especially over remote and inaccessible or large land areas. The RMSEs reported in this study were far less than that required for biomass estimation procedures in European countries, with an error not exceeding ±20% [22].

We also explored the use of 2D geospatial imagery not only for total plant biomass estimation but also for the estimation of wood and leaf-twig biomass components. Estimates of woody plant components have rarely been related to remotely sensed images [40]. Our results indicated that wood and leaf-twig biomass can be estimated reasonably well with canopy area and diameter values derived from aerial images of the resolution utilized in this study. Wood biomass has the highest energy content of the plant components and would likely be the most desirable portion of woody plants from a bioenergy perspective. Alternatively, leaf-twig biomass may be linked to the estimation of leaf area index (LAI), providing clues to estimate water use patterns and potential competitive effects with herbaceous understories. Leaves also conduct photosynthesis and are the primary energy exchange medium between plants and the environment, and are thus linked to productivity and biogeochemical cycles of a site [22,40-43].

The relationships developed between image measured canopy area and that measured on the site in this study were higher than that reported by [28]. These authors found that the QuickBird data accounted for 85% of the variation in ground measured eastern redcedar (*Juniperus virginiana*) biomass. A stronger relationship (r^2 = 0.98) compared to that shown here between image and ground measured eastern redcedar canopy area was reported by [28]. An identical relationship (r^2 = 0.92) between LIDAR and field measured biomass to that in this study was reported by [44], whereas a weaker relationship (r^2 = 0.62) was found by [45], who related LISS-IV estimated and field measured biomass. A good relationship (r^2 = 0.78) between Aleppo pine (*Pinus halepensis*) biomass measured in the field and estimated from a high-quality aerial photograph was reported in [46]. Image estimated biomass can serve as an alternative to field sampling in forest biomass research, given adequate temporal and spatial coverage of the study area [46].

Recently, there has been increasing amounts of base-line information required at large spatial scales as part of woody biomass operational management plans. Such details may include ecological diversity issues, partial harvesting (*i.e.*, leaving old and large trees), or patch harvesting for bioenergy, wildlife or hydrological concerns. These new requirements demand a greater level of detail in remote sensing data. Higher resolution data with individual tree-scale details allows for better precision in management decisions [1,3,40,47-49]. The individual plant approach offers a better understanding of the sources of uncertainty and errors associated with remote sensing estimates when predicting biomass at a large scale [48]. With the availability of high spatial and spectral resolution, multispectral and hyperspectral satellite and aerial remote sensing data in the recent times, it is now possible to estimate both individual tree and cluster or stand level biomass [28,40] at relatively low resource requirement. The equations derived and validated in this study are easy to comprehend and may be used with high confidence by managers, ecologists, scientists, and ranchers seeking rapid estimates of juniper biomass and possibly other tree species with similar growth forms for use in management and in ecological models.

Accurate estimates of biomass are also critical for monitoring carbon dynamics in rangelands, which is considered essential in assessing regional or global carbon budgets. Growing interest in quantifying ecosystem carbon stocks and/or potential bioenergy uses has triggered various research initiatives to explore possibilities of using non-destructive methods to estimate biomass [10, 28]. In particular, there has been a noticeable inclination towards using aerial or satellite remote sensing images to estimate biomass [3,28,50-53]. Regional companies are currently considering the use of eastern redcedar in gasification facilities for the production of steam or electricity for small rural communities, as well as in ethanol production, or for the production of woody pellets for residential heating [28]. With respect to potential biomass estimation for bioenergy uses, the aerial extent of the images will have to be sufficient to cover the large land areas (>50,000 ha) that are necessary for sustainable feedstock supply to a processing facility [28,54]. Such estimations may need to rely on 2D imagery, such as those from the National Agricultural Imagery Program (NAIP) of the USDA [35,53,55] or satellite images that can provide appreciable details at a regional scale and can be geospatially integrated with other features, such as road networks, waterways, etc. to facilitate harvest plans. Otherwise, if not impossible, it is too costly, labour intensive, and time consuming to estimate biomass via ground-based measurements [28].

Data reported worldwide has indicated that for most rangeland woody plants, the most accurate method of non-destructively estimating biomass is to measure basal stem diameter, crown diameter, plant height, and crown volume [7,56-61]. Basal stem diameter and diameter at breast height, however are not easily measurable parameters for juniper due to its highly compacted canopy and a high density of low growing stems restricting access to the central base stem [10]. Because of spherical growth forms and dense lateral growth, with a very compact canopy and a high density of low growing stems, these researchers determined that canopy area, determined by measuring canopy diameter in two directions, was the most suitable variable for predicting total tree AGM. Height was a less effective variable to predict biomass and including height to calculate canopy volume resulted in little improvement in the biomass estimation accuracy compared to canopy area alone. Since it is not possible to measure canopy height or stem diameter at breast height from 2D geospatial imagery, using canopy area and/or diameter measurements would appear to be the most suitable approach for predicting juniper tree biomass from 2D imagery. Imagery classification and converting classification to vector to obtain an automated canopy area of all trees in an area are easy and quick processes. The ability to remotely estimate biomass using canopy area would provide a major benefit for research and management efforts.

CONCLUSION

Though remote sensing data is becoming increasingly important in forest biomass estimation, little attention has been directed to rangeland woody biomass. There is an obvious need for large scale measurements to improve our understanding of the effects of ecological processes, such as woody encroachment into grasslands and the carbon contribution of rangeland woody plants to local or regional carbon stocks. The current study was undertaken to explore the relationship between canopy area and the diameter of individual red berry juniper trees determined via aerial images, and the biomass of the trees. A major finding of this research is that the method is highly automated and repeatable, and provides the ability to estimate and map juniper biomass using image classification in addition to manual digitizing. Results suggest that 2D geospatial imagery may be a powerful tool for estimating juniper biomass and in certain situations with adequate ground-based calibration curves, could replace ground measurement in many areas. Imagery classification of woody canopies can be a relatively straight forward process when the imagery is acquired during the dormant season when there is a strong spectral contrast between evergreen juniper and leafless deciduous trees and/or dormant herbaceous plants.

ACKNOWLEDGEMENTS

This research was funded in part by a grant from John Deere Renewables, Inc., and the Texas AgriLife Bioenergy Initiative Fund. Copper Breaks State Park and the Whatley Ranch, Crowell, TX provided harvest sites for this project. Images used in this study were made available by the USDA-NRCS. We also thank Mimi Roy for technical discussion and Andy Bell for assisting with data collection.

REFERENCES

1. N. F. Glenn, L. P. Spaete, T. T. Sankey, D. R. Derryberry, S. P. Hardegree and J. J. Mitchell, "Errors in LiDAR-Derived Shrub Height and Crown Area on Sloped Terrain," *Journal of Arid Environments*, Vol. 75, No. 4, 2011, pp. 377-382. doi:10.1016/j.jaridenv.2010.11.005
2. J. E. Nichol and M. L. R. Sarker, "Improved Biomass Estimation Using the Texture Parameters of Two High-Resolution Optical Sensors," *IEEE Transactions on Geoscience And Remote Sensing*, Vol. 49, No. 3, 2011, pp. 930-948. doi:10.1109/TGRS.2010.2068574_
3. E. K. Strand, L. A. Vierling, A. M. S. Smith and S. C. Bunting, "Net Changes in Aboveground Woody Carbon Stock in Western Juniper Woodlands, 1946-1998," *Journal of Geophysical Research D: Atmospheres*, Vol. 113, No. G01013, 2008, pp. 1-13.
4. Z. Wang, C. B. Schaaf, P. Lewis, Y. Knyazikhin, M. A. Schull, A. H. Strahler, T. Yao, R. B. Myneni, M. J. Chopping and B. J. Blair, "Retrieval of Canopy Height Using Moderate-Resolution Imaging Spectroradiometer (MODIS) Data," *Remote Sensing of Environment*, Vol.
5. 115, No. 6, 2011, pp. 1595-1601. doi:10.1016/j.rse.2011.02.010_
6. K. Brinkmann, U. Dickhoefer, E. Schlecht and A. Buerkert, "Quantification of Aboveground Rangeland Productivity and Anthropogenic Degradation on the Arabian Peninsula Using Landsat Imagery and Field Inventory Data," *Remote Sensing of Environment*, Vol. 115, No. 2, 2011, pp. 465-474. doi:10.1016/j.rse.2010.09.016_
7. C. Huang, S. E. Marsh, M. P. McClaran and S. R. Archer, "Postfire Stand Structure in a Semiarid Savanna: Cross- Scale Challenges Estimating Biomass," *Ecological Applications*, Vol. 17, No. 7, 2007, pp. 1899-1910. doi:10.1890/06-1968.1_
8. E. T. A. Mitchard, S. S. Saatchi, S. L. Lewis, T. R. Feldpausch, I. H. Woodhouse, B. Sonké, C. Rowland and P. Meir, "Measuring Biomass Changes Due to Woody Encroachment and Deforestation/Degradation in a Forest- Savanna Boundary Region of Central Africa Using Multi- Temporal L-Band Radar Backscatter," *Remote Sensing of Environment*, Vol. 115, No. 11, 2011, pp. 2861-2873. doi:10.1016/j.rse.2010.02.022_
9. K. P. Paudel and P. Andersen, "Assessing Rangeland Degradation Using Multi Temporal Satellite Images and Grazing Pressure Surface Model in Upper Mustang, Trans Himalaya, Nepal," *Remote Sensing of Environment*, Vol. 114, No. 8, 2010, pp. 1845-1855. doi:10.1016/j.rse.2010.03.011
10. R. Sonnenschein, T. Kuemmerle, T. Udelhoven, M. Stellmes and P. Hostert, "Differences in Landsat-Based Trend Analyses in Drylands Due to the Choice of

Vegetation Estimate," *Remote Sensing of Environment*, Vol. 115, No. 6, 2011, pp. 1408-1420. doi:10.1016/j.rse.2011.01.021

11. R. J. Ansley, M. Mirik, B. W. Surber and S. C. Park, "Canopy Area and Aboveground Mass of Individual Redberry Juniper (*Juniperus pinchotii* Sudw.) Trees," *Rangeland Ecology & Management*, Vol. 65, No. 2, 2012, pp. 189-195. doi:10.2111/REM-D-11-00112.1
12. M. A. Alrababah, M. N. Alhamad, A. L. Bataineh, M. M. Bataineh and A. F. Suwaileh, "Estimating East Mediterranean Forest Parameters Using Landsat ETM," *International Journal of Remote Sensing*, Vol. 32, No. 6, 2011, pp. 1561-1574. doi:10.1080/01431160903573235
13. M. L. Clark, D. A. Roberts, J. J. Ewel and D. B. Clark, "Estimation of Tropical Rain Forest Aboveground Biomass with Small-Footprint Lidar and Hyperspectral Sensors," *Remote Sensing of Environment*, Vol. 115, No. 11, 2011, pp. 2931-2942. doi:10.1016/j.rse.2010.08.029
14. S. Englhart, V. Keuck and F. Siegert, "Aboveground Biomass Retrieval in Tropical Forests—The Potential of Combined X- and L-Band SAR Data Use," *Remote Sensing of Environment*, Vol. 115, No. 5, 2011, pp. 1260- 1271. doi:10.1016/j.rse.2011.01.008
15. M. T. Gebreslasie, F. B. Admed, J. A. N. Van Aardt and F. Blakeway, "Individual Tree Detection Based on Variable and Fixed Window Size Local Maximum Filtering Applied to IKONOS Imagery for Even-Aged *Eucalyptus* Plantation Forest," *International Journal of Remote Sensing*, Vol. 32, No. 15, 2011, pp. 4141-4154. doi:10.1080/01431161003777205
16. Y. Ke and L. J. Quackenbush, "A Review of Methods for Automatic Individual Tree-Crown Detection and Delineation from Passive Remote Sensing," *International Journal of Remote Sensing*, Vol. 32, No. 17, 2011, pp. 4725- 4747. doi:10.1080/01431161.2010.494184
17. S. C. Popescu, K. Zhao, A. Neuenschwander and C. Lin, "Satellite Lidar vs. Small Footprint Airborne Lidar: Comparing the Accuracy of Aboveground Biomass Estimates and Forest Structure Metrics at Footprint Level," *Remote Sensing of Environment*, Vol. 115, No. 11, 2011, pp. 2786- 2797. doi:10.1016/j.rse.2011.01.026
18. L. R. Sarker and J. E. Nichol, "Improved Forest Biomass Estimates Using ALOS AVNIR-2 Texture Indices," *Remote Sensing of Environment*, Vol. 115, No. 4, 2011, pp. 968-977. doi:10.1016/j.rse.2010.11.010
19. C. Yang, H. Huang and S. Wang, "Estimation of Tropical Forest Biomass Using Landsat TM Imagery and Permanent Plot Data in Xishuangbanna, China," *International Journal of Remote Sensing*, Vol. 32, No. 20, 2011, pp. 5741-5756. doi:10.1080/01431161.2010.507677
20. E. Næsset, "Estimating Above-Ground Biomass in Young Forests with Airborne Laser Scanning," *International Journal of Remote Sensing*, Vol. 32, No. 2, 2011, pp. 473-501. doi:10.1080/01431160903474970

21. G. Sun, K. J. Ranson, Z. Guo, Z. Zhang, P. Montesano and D. Kimes, "Forest Biomass Mapping from Lidar and Radar Synergies," *Remote Sensing of Environment*, Vol. 115, No. 11, 2011, pp. 2906-2916. doi:10.1016/j.rse.2011.03.021

22. G. Chen and G. J. Hay, "An Airborne Lidar Sampling Strategy to Model Forest Canopy Height from Quickbird Imagery and GEOBIA," *Remote Sensing of Environment*, Vol. 115, No. 6, 2011, pp. 1532-1542. doi:10.1016/j.rse.2011.02.012

23. T. Le Toan, S. Quegan, M. W. J. Davidson, H. Balzter, P. Paillou, K. Papathanassiou, S. Plummer, F. Rocca, S. Saatchi, H. Shugart and L. Ulander, "The BIOMASS Mission: Mapping Global Forest Biomass to Better Understand the Terrestrial Carbon Cycle," *Remote Sensing of Environment*, Vol. 115, No. 11, 2011, pp. 2850-2860. doi:10.1016/j.rse.2011.03.020

24. Swatantran, R. Dubayah, D. Roberts, M. Hofton and J. B. Blair, "Mapping Biomass and Stress in the Sierra Nevada Using Lidar and Hyperspectral Data Fusion," *Remote Sensing of Environment*, Vol. 115, No. 11, 2011, pp. 2917-2930. doi:10.1016/j.rse.2010.08.027

25. Song, M. B. Dickinson, L. Su, S. Zhang and D. Yaussey, "Estimating Average Tree Crown Size Using Spatial Information from Ikonos and QuickBird Images: Across- Sensor and Across-Site Comparisons," *Remote Sensing of Environment*, Vol. 114, No. 5, 2010, pp. 1099-1107. doi:10.1016/j.rse.2009.12.022

26. P. Gonzalez, G. P. Asner, J. J. Battles, M. A. Lefsky, K. M. Waring and M. Palace, "Forest Carbon Densities and Uncertainties from Lidar, QuickBird, and Field Measurements in California," *Remote Sensing of Environment*, Vol. 114, No. 7, 2010, pp. 1561-1575. doi:10.1016/j.rse.2010.02.011

27. R. J. Ansley, H. T. Weidemann, M. J. Castellano and J. E. Slosser, "Herbaceous Restoration of Juniper-Dominated Grasslands with Chaining and Fire," *Rangeland Ecology & Management*, Vol. 59, No. 2, 2006, pp. 171-178. doi:10.2111/05-095R1.1

28. Ozdemir, "Estimating Stem Volume by Tree Crown Area and Tree Shadow Area Extracted from Pan-Sharpened QuickBird Imagery in Open Crimean Juniper Forests," *International Journal of Remote Sensing*, Vol. 29, No. 19, 2008, pp. 5643-5655. doi:10.1080/01431160802082155

29. P. J. Starks, B. C. Venuto, J. A. Eckroat and T. Lucas, "Measuring Eastern Redcedar (*Juniperus virginiana* L.) Mass with the Use of Satellite Imagery," *Rangeland Ecology & Management*, Vol. 64, No. 2, 2011, pp. 178-186. doi:10.2111/REM-D-10-00057.1

30. National Oceanic and Atmospheric Administration, Climate Data Center, Asheville, 2006. http://www.ncdc.noaa.gov/ oa/ncdc.html

31. NRCS, United States Department of Agriculture, Natural Resource Conservation Service, Plants Database Website, 2011. http://plants.usda.gov

32. NRCS, United States Department of Agriculture, Natural Resource Conservation Service, Soil Series Descriptions Website, 2011. http://soils.usda.gov

33. H. Everitt, C. Yang and H. B. Johnson, "Canopy Spectra and Remote Sensing of Ashe Juniper and Associated Vegetation," *Environmental Monitoring and Assessment*, Vol. 130, No. 1-3, 2007, pp. 403-413. doi:10.1007/s10661-006-9407-2

34. C. Yang, J. H. Everitt and H. B. Johnson, "Applying Image Transformation and Classification Techniques to Airborne Hyperspectral Imagery for Mapping Ashe Juniper Infestations," *International Journal of Remote Sensing*, Vol. 30, No. 11, 2009, pp. 2741-2758. doi:10.1080/01431160802555812

35. B. Somers, G. P. Asner, L. Tits and P. Coppin, "Endmember Variability in Spectral Mixture Analysis: A Review," *Remote Sensing of Environment*, Vol. 115, No. 7, 2011, pp. 1603-1616. doi:10.1016/j.rse.2011.03.003

36. W. Davies, S. L. Petersen, D. D. Johnson, D. B. Davis, M. D. Madsen, D. L. Zvirzdin and J. D. Bates, "Estimating Juniper Cover from National Agriculture Imagery Program (NAIP) Imagery and Evaluating Relationships between Potential Cover and Environmental Variables," *Rangeland Ecology & Management*, Vol. 63, No. 6, 2010, pp. 630-637. doi:10.2111/REM-D-09-00129.1

37. S. L. Petersen and T. K. Stringham, "Development of GIS-Based Models to Predict Plant Community Structure in Relation to Western Juniper Establishment," *Forest Ecology and Management*, Vol. 256, No. 5, 2008, pp. 981-989. doi:10.1016/j.foreco.2008.05.058

38. T. T. Sankey, N. Glenn, S. Ehinger, A. Boehm and S. Hardegree, "Characterizing Western Juniper Expansion via a Fusion of Landsat 5 Thematic Mapper and Lidar data," *Rangeland Ecology & Management*, Vol. 63, No. 5, 2010, pp. 514-523. doi:10.2111/REM-D-09-00181.1

39. T. T. Sankey and M. J. Germino, "Assessment of Juniper Encroachment with the Use of Satellite Imagery and Geospatial Data," *Rangeland Ecology & Management*, Vol. 61, No. 4, 2008, pp. 412-418. doi:10.2111/07-141.1

40. Hufkens, J. Bogaert, Q. H. Dong, L. Lu, C. L. Huang, M. G. Ma, T. Che, X. Li, F. Veroustraete and R. Ceulemans, "Impacts and Uncertainties of Upscaling of Remote-Sensing Data Validation for a Semi-Arid Woodland," *Journal of Arid Environments*, Vol. 72, No. 8, 2008, pp. 1490-1505. doi:10.1016/j.jaridenv.2008.02.012

41. R. M. Lucas, A. C. Lee and P. J. Bunting, "Retrieving Forest Biomass through Integration of CASI and LIDAR Data," *International Journal of Remote Sensing*, Vol. 29, No. 5, 2008, pp. 1553-1577. doi:10.1080/01431160701736497

42. D. F. Levia Jr., "A Generalized Allometric Equation to Predict Foliar Dry Weight on the Basis of Trunk Diameter for Eastern White Pine (*Pinus strobus* L.)," *Forest Ecology and Management*, Vol. 255, No. 5-6, 2008, pp. 1789-1792. doi:10.1016/j.foreco.2007.12.001

43. S. D. Roberts, T. J. Dean, D. L. Evans, J. W. McCombs, R. L. Harrington and P. A. Glass, "Estimating Individual Tree Leaf Area in Loblolly Pine Plantations Using LiDAR-Derived Measurements of Height and Crown Dimensions," *Forest*

Ecology and Management, Vol. 213, No. 1-3, 2005, pp. 54-70. doi:10.1016/j.foreco.2005.03.025

44. Wang, J. E. R. Hunt, J. J. Qu, X. Hao and C. S. T. Daughtry, "Towards Estimation of Canopy Foliar Biomass with Spectral Reflectance Measurements," *Remote Sensing of Environment*, Vol. 115, No. 3, 2011, pp. 836- 840. doi:10.1016/j.rse.2010.11.011
45. R. M. Lucas, N. Cronin, M. Moghaddam, A. Lee, J. Armston, P. Bunting and C. Witte, "Integration of Radar and Landsat Derived Foliage Projected Cover for Woody Regrowth Mapping, Queensland, Australia," *Remote Sensing of Environment*, Vol. 100, No. 3, 2006, pp. 388-406. doi:10.1016/j.rse.2005.09.020
46. R. Madugundu, V. Nizalapur and C. C. Jha, "Estimation of LAI and Above-Ground Biomass in Deciduous Forests: Western Ghats of Karnataka, India," *International Journal of Applied Earth Observation and Geoinformation*, Vol. 10, No. 2, 2008, pp. 211-219. doi:10.1016/j.jag.2007.11.004
47. B. Massada, Y. Carmel, G. E. Tzur, J. M. Grünzweig and D. Yakir, "Assessment of Temporal Changes in Aboveground Forest Tree Biomass Using Aerial Photographs and Allometric Equations," *Canadian Journal of Forestry Research*, Vol. 36, No. 10, 2006, pp. 2585-2594. doi:10.1139/x06-152
48. H. Holmström, "Estimation of Single-Tree Characteristics Using the *k*NN Method and Plotwise Aerial Photograph Interpretations," *Forest Ecology and Management*, Vol. 167, No. 1-3, 2002, pp. 303-314. doi:10.1016/S0378-1127(01)00720-4
49. S. C. Popescu, "Estimating Biomass of Individual Pine Trees Using Airborne Lidar," *Biomass and Bioenergy*, Vol. 31, No. 9, 2007, pp. 646-655. doi:10.1016/j.biombioe.2007.06.022
50. G. Whiteman and J. R. Brown, "Assessment of a Method for Mapping Woody Plant Density in a Grassland Metrix," *Journal of Arid Environments*, Vol. 38, No. 2, 1998, pp. 269-282. doi:10.1006/jare.1997.0325
51. G. P. Asner, S. Archer, R. F. Hughes, R. J. Ansley and C. A. Wessman, "Net Changes in Regional Woody Vegetation Cover and Carbon Storage in Texas Drylands, 1937- 1999," *Global Change Biology*, Vol. 9, No. 3, 2003, pp. 316-335. doi:10.1046/j.1365-2486.2003.00594.x
52. C. Huang, G. P. Asner, R. E. Martin, N. N. Barger and J. C. Neff, "Multiscale Analysis of Tree Cover and Above-ground Carbon Stocks in Pinyon-Juniper Woodlands," *Ecological Applications*, Vol. 19, No. 3, 2009, pp. 668- 681. doi:10.1890/07-2103.1
53. Masera, A. Ghilardi, R. Drigo and M. Angel Trossero, "WISDOM: A GIS-Based Supply Demand Mapping Tool for Woodfuel Management," *Biomass and Bioenergy*, Vol. 30, No. 7, 2006, pp. 618-637. doi:10.1016/j.biombioe.2006.01.006
54. M. Mirik and R. J. Ansley, "Comparison of Ground-measured and Image-Classified Honey Mesquite (*Prosopis glandulosa*) Canopy Cover in Texas," *Rangeland Ecology & Management*, Vol. 65, No. 1, 2012, pp. 85-95. doi:10.2111/REM-D-11-00073.1

55. R. J. Ansley, M. Mirik and M. J. Castellano, "Structural Biomass Partitioning in Regrowth and Undisturbed Mesquite (*Prosopis glandulosa*): Implications for Bioenergy Uses," *Global Change Biology Bioenergy*, Vol. 2, No. 1, 2010, pp. 26-36. doi:10.1111/j.1757-1707.2010.01036.x
56. E. Strand, A. Smith, S. Bunting, L. Vierling, D. Hann and P. Gessler, "Wavelet Estimation of Plant Spatial Patterns in Multitemporal Aerial Photography," *International Journal of Remote Sensing*, Vol. 27, No. 10, 2006, pp. 2049- 2054. doi:10.1080/01431160500444764
57. E. N. Chidumayo, "Above-Ground Woody Biomass Structure and Productivity in a Zambezian Woodland," *Forest Ecology and Management*, Vol. 36, No. 1, 1990, pp. 33-46. doi:10.1016/0378-1127(90)90062-G
58. Lufafa, I. Diédhiou, N. A. S. Ndiaye, M. Séné, F. Kizito, R. P. Dick and J. S. Noller, "Allometric Relationships and Peak-Season Community Biomass Stocks of Native Shrubs in Senegal's Peanut Basin," *Journal of Arid Environments*, Vol. 73, No. 3, 2009, pp. 260-266. doi:10.1016/j.jaridenv.2008.09.020
59. K. Northup, S. F. Zitzer, S. Archer, C. R. McMurtry and T. W. Boutton, "Above-Ground Biomass and Carbon and Nitrogen Content of Woody Species in a Subtropical Thornscrub Parkland," *Journal of Arid Environments*, Vol. 62, No. 1, 2005, pp. 23-43. doi:10.1016/j.jaridenv.2004.09.019
60. L. F. Ohmann, D. F. Grigal and R. B. Brander, "Biomass Estimation for Five Shrubs from Northeastern Minnesota," Research Paper NC-133, US Department of Agriculture, Forest Service, North Central Forest Experiment Station, St. Paul, 1976.
61. Rosenschein, T. Tietema and D. O. Hall, "Biomass Measurement and Monitoring of Trees and Shrubs in a Semi-Arid Region of Central Kenya," *Journal of Arid Environments*, Vol. 42, No. 2, 1999, pp. 97-116. doi:10.1006/jare.1999.0509
62. T. Tietema, "Biomass Determination of Fuelwood Trees and Bushes of Botswana, Southern Africa," *Forest Ecology and Management*, Vol. 60, No. 3-4, 1993, pp. 257- 269. doi:10.1016/0378-1127(93)90083-Y

9
CHAPTER

USING SPOT AND AERIAL FALSE-COLOR INFRARED (FCIR) IMAGERY

Alvan Karlin, Mark Fulkerson and Gene Altman

Keywords: Storm water modelling, Tropical Storm Debby, model verification, high water marks, aerial imagery

INTRODUCTION

Floods are a natural phenomenon that cause an average of approximately $6 billion in damages annually (Sarmiento and Miller [1]). In some cases, flooding is expected, however often the public is unaware of the risk and unprepared for the consequences. For decades, local, state and federal agencies have worked to develop tools to accurately identify flood risk and to quickly respond to flooding when it happens. Flood protection is one of the four major missions of the Southwest Florida Water Management District (SWFWMD). Within this mission, the SWFWMD cooperates with federal and state entities, and local governments to develop the Federal Emergency Management Agency (FEMA) Digital Flood Insurance Rate Maps (DFIRMs). During the development of the DFIRMs, the SWFWMD investigates:

a. the results of storm water models;
b. a database of historical photographs and high water elevations; and
c. information from local residents.

One of the more important aspects of the storm water models (sometimes referred to as Hydrologic and Hydraulic (H & H) models), is the model verification simulation

Citation: Karlin A., Fulkerson M., Altman, G., Using SPOT and Aerial False-Color Infrared (FCIR) Imagery, doi:10.3390/geosciences6020024.

for a given model. During model verification, rainfall depths from an actual storm event are simulated to produce floodplain results, which are then compared to documented high water information associated with that same event. Results that reasonably match the high water information provide a high level of confidence in model quality, thereby assuring a high level of accuracy when modelling design storms, such as the 100-year flood.

Singhofen [2] gives an excellent background on storm water model verification, focusing on empirical methodologies. Sawaya et al. [3] demonstrated applications of IKONOS-2 and QuickBird satellite imagery for mapping lake water levels and impervious surface determination. Although their focus is at the local scale, they [3] do not apply the analysis to model verification. Miller et al. [4] used MODIS (Moderate Resolution Imaging Spectrometer) satellite data to verify sediment transport models and found that higher resolution imagery provides better verification results compared to lower resolution AVHRR (Advanced Very High Resolution Radiometer) or SeaWiFS (Sea-Viewing Wide Field-of-View Sensor) imagery. Haq et al. [5] used remotely-sensed satellite imagery to monitor flooding and provide damage assessments. Others (Jung et al. [6]) have used two-dimensional (2D) satellite imagery to validate surface water heights and slopes derived from 2D hydrodynamic models. Their results compared well to monitored surface water levels, however, they were limited to the response of a channelized tributary, and have not been tested on the complex dynamics of an entire watershed that includes both above and below ground flowpaths. More recently, Tarpanelle et al. [7] used Synthetic Aperture Radar (SAR) imagery to verify Manning's Roughness coefficients for a channel and floodplain, and Kunyeun et al. [8] used flood extent data for model verification of underground storm sewers in an urban area. Remote sensing has been used extensively for mapping purposes, however, its application to model verification is limited. This study uses remotely-sensed images from a recent flood event to test and verify floodplain model results for a watershed in west central Florida.

STUDY AREA

The Peck Sink Watershed is approximately 17 square miles within Hernando County, Florida (Figure 1). Its topography ranges in elevation from 20 to 250 feet (NAVD88), which is unique when compared to the generally flat landscape of Florida. With a mix of hardwoods, cropland and low-density residential areas, this watershed is dominated by poorly drained soils and sinkhole features that drain the land. This combination of topographic relief and clay soils near land surface has led to a history of flooding.

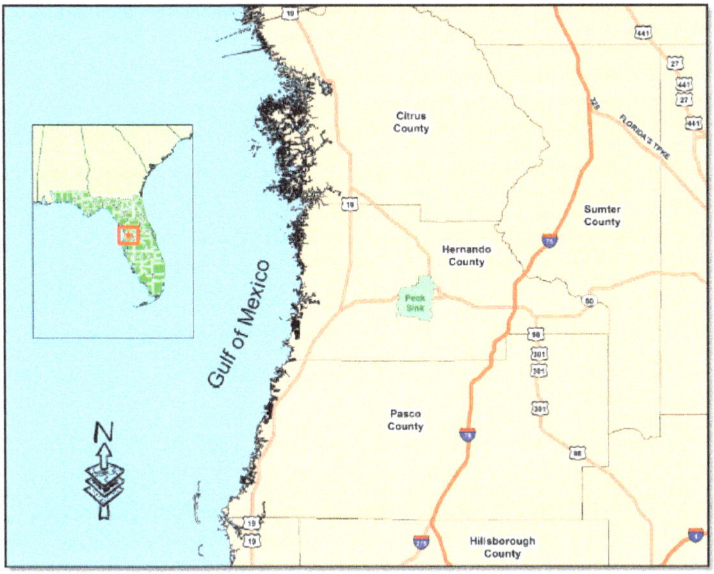

Figure 1. Location of the Peck Sink Watershed in Hernando County, Florida.

MATERIALS AND METHODS

During the summer of 2012, while the SWFWMD was conducting a study of the Peck Sink Watershed to better understand the watershed's flood risk, record rainfall occurred from Tropical Storm Debby. This storm event developed in the central Gulf of Mexico on Saturday, 23 June 2012, approached and stalled over southwestern Florida on Sunday, 24 June 2012, before moving north-westward and making landfall over Steinhatchee, Florida on 26 June 2012. Although the maximum winds (95 km/h) and minimum pressures (990 mb) were recorded to the northwest of the SWFWMD, the storm delivered over 13 inches of rainfall as it passed over parts of the 16-county area of the SWFWMD. Citrus, Hernando, and Pasco Counties experienced rainfall during this time period that would normally be expected only once in a 100-year period, *i.e.*, a storm with a 1% chance of occurrence in any given year. Hence offered the SWFWMD had an opportunity to test the floodplain model that was being developed for the Peck Sink Watershed. Remotely-sensed images were obtained and compared to modelled floodplains

Storm Water Modelling

Storm water modelling and flood mapping for the Peck Sink Watershed, Hernando County, Florida was conducted by CH2M Hill Engineers, Inc. (Tampa, FL, USA) and presented to the SWFWMD on February 2016. The model was based on LiDAR-elevations and breaklines collected in 2007. The LiDAR data, with 1 m postings and a fundamental vertical accuracy of 9.25 cm, was collected by Woolpert, Inc. (Dayton,

OH, USA) and processed into hydro-enforced and hydro-flattened Digital Elevation Models (DEMs) with a 5-foot × 5-foot cell size. The DEMs were distributed to CH2M Hill Engineers, Inc., to assist in the development of the floodplain model. CH2M Hill Engineers, Inc. used the best "engineering judgement" and standards of practice to augment the "surface DEM" with below-surface features, such as pipes, drop structures, and weirs, to direct water flows. For the Peck Sink Watershed, the Interconnected Channel and Pond Routing model (ICPRv3; Streamline Technologies, Orlando, FL, USA) was used to simulate flooding for several design storms and actual events, including Doppler rainfall from Tropical Storm Debby in June 2012. Results from the model were converted to Esri-shapefiles and subsequently stored in either an Esri file-geodatabase or the SWFWMD Enterprise geodatabase.

Image Acquisition

When a "state of emergency" for Florida was declared on 25 June 2012, the state qualified for Federal Emergency Management Agency (FEMA) assistance. A portion of that assistance included the acquisition of satellite images (SPOT, Satellite Pour l'Observation del la Terra) when the satellite passed overhead on 27 June 2012, approximately 48 h after the storm started, moving northward and out of the SWFWMD. The SPOT multi-spectral image for the north-central portion of the SWFWMD (SP 04N28_475187W082_3623332012062800000000MS00_GG003002001.tif) was of particular interest, as it focused on Hernando County, Florida, one of the areas significantly affected by flooding.

Although the SPOT data was obtained within a 48-h window of Tropical Storm Debby, there were concerns that the 20 m Ground Sample Distance (GSD) and the Infrared bandwidth frequency would be too coarse to accurately map the extent of the flooding. The SWFWMD contracted a local aerial firm (Aerial Cartographics of America, Orlando, FL, USA) to obtain a high resolution false-colour infrared (fCIR) digital image within specific portions of Citrus, Hernando and Pasco Counties, Florida that received the highest rainfall depths. The image, acquired with a Vexcel UltraCamX digital aerial sensor using 60% forward and 40% sidelap, produced a 9-inch raw pixel GSD. Flights were conducted between 2:00 p.m. and 4:00 p.m. on 27 June 2012, and concluded, the next day, 28 June 2012 between 10:00 a.m. and 1:00 p.m. Photography was taken at a 30° sun angle and was free of smoke and clouds. Final project deliverables were referenced to Florida State Plane West, NAD83 (2007) U.S. Survey feet and tiled into 5000-foot × 5000-foot tiles. The SWFWMD supplied a 5-foot LiDAR-derived DEM referenced to the Florida State Plane West, NAD 2007, NAVD 1988, U.S. Survey feet for ortho-rectification. All work was performed under the supervision of a licensed Florida Professional Surveyor and Mapper. Orthophotos were certified to meet or exceed a verified horizontal accuracy of 9.12 foot at the 95% confidence interval (3.0 feet Root Mean Square Error), as specified in the American Society of Photogrammetry and Remote Sensing (ASPRS) standards. No correction for building lean was made.

Model Verification

The presence of water shown on the remotely-sensed images was directly compared to the simulated flooding from the model using a combination of visual inspection and image analysis methods. The 4-band (RGBi) image was first examined for trends in spectral response to determine threshold values for waterbodies. After threshold values were selected, an extract by mask was run on the Near-Infrared (NIR) band, using Digital Number (DN) values of 75 and below. The resulting extracted raster was then reclassified from 1 to 5, with 1 being almost certainly water, and 5 being more likely shadowed land. The reclassified image was used alongside the original to identify flooding within the study area. High water elevations and flood photos from Tropical Storm Debby were obtained from residents, local governments and SWFWMD staff to document the flooding and assist with model verification.

RESULTS

Modelled floodplain results show that 21% of the total Peck Sink Watershed was inundated during Tropical Storm Debby. Both SPOT and fCIR images were captured within two to three days of the storm and compared to the simulated floodplain. Figure 2 shows the modelled Tropical Storm Debby floodplain overlain onto the SPOT image with water accumulation depicted in black. A large-scale visual comparison of the entire watershed showed reasonable identification of flooding in the central and southern portions of the watershed, where the landscape is flatter and lower in elevation. In the northern portion of the watershed, where the terrain is steeper and higher in elevation, the SPOT imagery showed little or no evidence of the flooding that was simulated by the model.

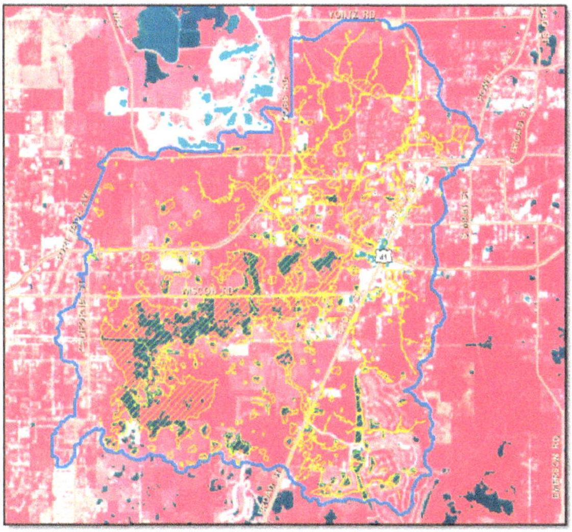

Figure 2. The Peck Sink Watershed (blue outline) showing Tropical Storm Debby modelled floodplain (yellow hatching) overlain onto the SPOT image (scale ~ 1:65,000).

The fCIR aerial image captured approximately seven square miles of the central and southern portions of the Peck Sink Watershed, three days after Tropical Storm Debby. A comparison of the SPOT and fCIR images (Figure 3) show an area in the central portion of the Peck Sink Watershed that experienced flooding. The deeper red signatures are interpreted to represent denser vegetation, and although the fCIR image more clearly identifies the flooding, neither image shows the entire presence of flooding in these obscured areas. Field-measured high water marks from Tropical Storm Debby were obtained at many locations, as shown in Figure 3, verifying flooding in some areas that were not visible in the remotely-sensed images.

A closer examination of a house that was completely surrounded by water from Tropical Storm Debby is shown in Figure 4. This figure compares the flood inundation visible in both the SPOT and fCIR imagery to the modelled+floodplain. In the SPOT image, probably as a result of the 20 m GSD, the house is not visible, although there are dark pixels, indicating water, in the vicinity. Given that there may be a slight horizontal displacement as a result of the geo-rectification process, the SPOT image does place the house (the brightest pixel) within the yellow circle. This image does a reasonably good job at identifying the flooding extent, however the abundance of non-black pixels indicates upland areas that are not present. The higher resolution fCIR image shows the floodwaters surrounding the house more clearly, although the modelled floodplain (yellow hatching) indicates the water should have been closer to the house at the peak of the flood.

An oblique image captured by the helicopter shortly after the fCIR aerial image, shows the same house surrounded by water on 29 June 2012 (Figure 5). A debris line, clearly visible in the image, indicates the peak flood level was closer to the house immediately following the storm event but had receded by the time the remotely-sensed image was captured. High water elevations, surveyed using this debris line as an indicator from the oblique image, closely matched the modelled floodplain elevations for Tropical Storm Debby in this vicinity.

The results of the image analysis provide additional insight into the use of remote sensing to verify model results. As expected, this analysis clearly identified water in open areas, although there were difficulties in identifying water in areas obscured by dense vegetation (Figure 6). In many cases, a visual inspection of the DEM was necessary to confirm the modelled floodplain results in these obscured areas. Figure 7 shows the modelled floodplain from Tropical Storm Debby as compared to both the DEM and the results of the image analysis. As expected, the floodplain matches well with the contour of the land surface elevation from the DEM. Although the entire floodplain was not identified by the image analysis, the resulting raster fragments extend to the modelled boundary, indicating the overall extent of flooding.

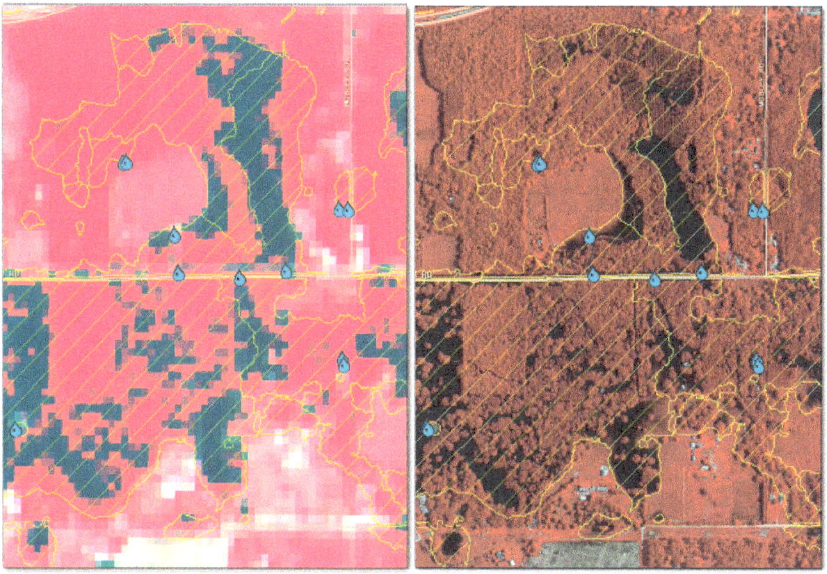

Figure 3. SPOT image (**left**) and fCIR image (**right**) with Tropical Storm Debby modelled floodplain polygons (yellow hatching). Field-measured high water marks are depicted by blue raindrops (scale ~ 1:12,500).

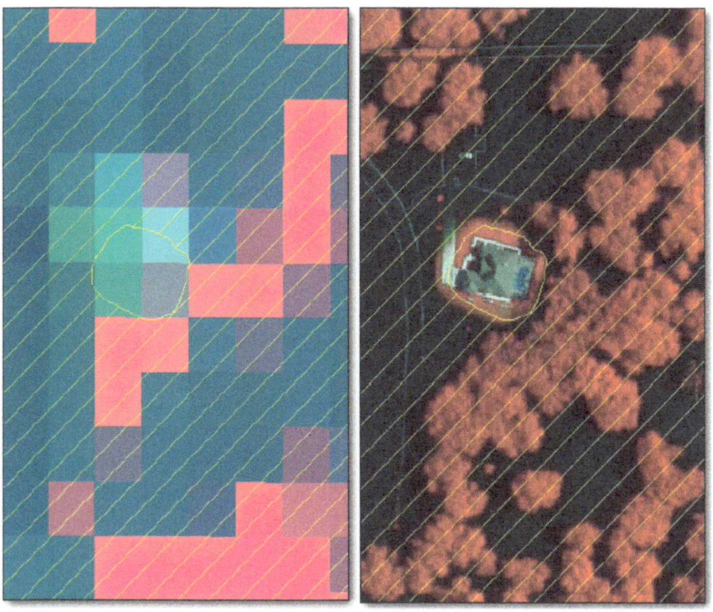

Figure 4. SPOT image (**left**) and aerial fCIR image (**right**) with modelled floodplain (yellow hatching) surrounding the house (scale ~ 1:1500).

Figure 5. Oblique image of house taken during Tropical Storm Debby flooding event (June 2012; care of Hernando County Emergency Management). scale ~ 1:3000).

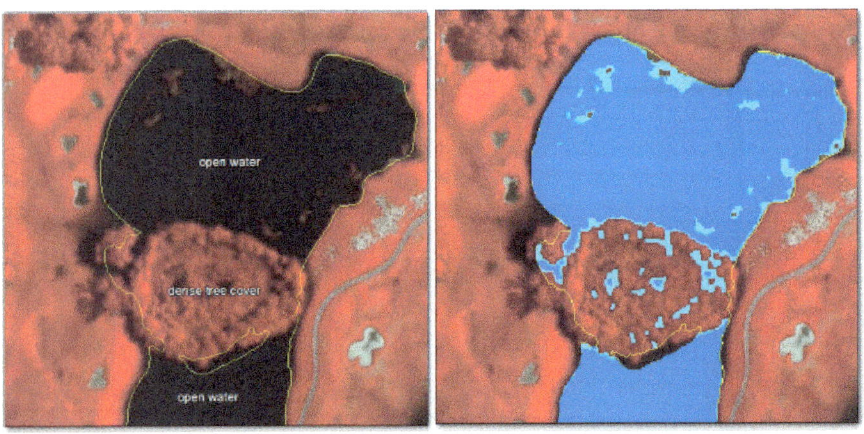

Figure 6. Image analysis results (**right**, indicated in blue) for the fCIR image (**left**) in obscured and unobscured areas as compared to Tropical Storm Debby modelled floodplain (yellow outline) (scale ~ 1:3000).

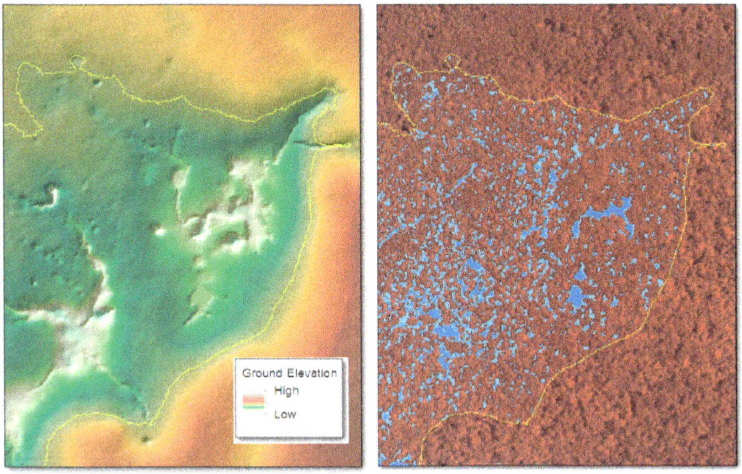

Figure 7. Comparison of Tropical Storm Debby modelled floodplain (yellow outline) to the DEM (**left**) and results (**right**, indicated in blue) of the fCIR image analysis (scale ~ 1:10,000).

DISCUSSION AND CONCLUSIONS

Since the early 2000s, the SWFWMD's Watershed Management Program has been engaged in hydrologic and hydraulic modelling of watersheds throughout their 16 county service area. This program is designed to construct detailed floodplain models capable of identifying flood risk at the individual parcel level and to assist communities and local governments in the development of best management practices for flood protection. These goals require the floodplain models to include detailed information for both the conveyance ways above ground and complex pipe networks below ground.

The topology and hydrology of southwestern Florida are unlike many portions of the United States, where basins, sub-basins and catchments are clearly defined by flow ways, rivers, streams and creeks, forming dendritic drainage patterns. In contrast, the basins (catchments) in the SWFWMD are defined by a hierarchical nesting of sinks-low points in the terrain, and ridges-high points in the terrain. These form a deranged network of flow, where catchments fill and overflow, through spill points and low points along the ridges, between catchments and into adjacent catchments. In such deranged terrains, elevation differences as small as 0.5 foot can result in changing overflow directions and different flooding patterns. This sensitivity requires a characterization of the terrain that is highly resolute, accurate, and detailed. DEMs used by the SWFWMD are LiDAR-derived (Light Detection and Ranging) DEMs with cell sizes typically ranging between 2.5 and five feet and elevation accuracies in the 0.2 to 0.3-foot range (NAVD88; North American Vertical Datum of 1988). In addition to surface flows, commercial- and residential-developed watersheds also include extensive below ground drainage networks to route storm water in man-made flow patterns. Detailed measurements of outfall structures, underground pipe systems and discharge locations are also included into these floodplain models to accurately simulate the direction and quantity of storm water flows.

This detailed level of modelling is in sharp contrast to similar studies (Mason et al. [9], Khan et al. [10] and Karim et al. [11]) that used H & H modelling at smaller map scales with lower resolution imagery and topographic information. Mason et al. [9] compared flood inundation along the Thanes River (Oxford, UK) to remotely-sensed data, with a grid cell size of 50 m. Kahn et al. [10] used a similar methodology to construct a flood inundation model for the Nzoia River basin (a sub-basin of Lake Victoria in Africa). This model was constructed at a very small mapping scale using remotely sensed data with 250 m to 500 m resolution and SRTM (Shuttle Radar Topography Mission) 30 m post-spacing elevation data. More recently, Karim et al. [11] used 30 m SRTM data for elevation and MODIS (Moderate Resolution Image Spectrometer) to derive flood inundation maps for a riverine reach along the Fitzroy River in Australia. Hence, our study, conducted at a very large mapping scale, over deranged terrains with extensive below-surface, man-made alternations, provides new insights for the use of imagery for H & H model verification.

The development of accurate floodplain information for the Peck Sink Watershed required detailed topographic information coupled with a complete inventory of the below ground drainage network. The results of this study indicate that a combination of remotely-sensed imagery and field-measured high water marks can be used to verify floodplain model results for large-scale, highly-detailed, deranged watersheds. Although areas of flood inundation were reasonably identifiable in the SPOT imagery, the fine grained detail was obscured, limiting its ability to verify the modelled floodplain from Tropical Storm Debby. The detail obtained from the higher resolution imagery are in good agreement with the modelled floodplains, however, the overall extent of flooding was underestimated, as shown by the results of the image analysis. This limiting factor was mostly due to the inability of the fCIR image to clearly identify inundation in densely vegetated areas, which are highly characteristic of the Peck Sink Watershed. In addition, field data suggest that the floodwaters had receded in many areas before the remotely-sensed images were captured. These limitations required the use of additional tools including the DEM and field-measured high water marks to confirm the validity of the modelled floodplain results. This work shows how flood inundation can be inferred in areas not visible from remotely-sensed imagery. Future work might include developing a methodology to more accurately identify flood inundation in areas obscured by vegetation.

Flood events such as Tropical Storm Debby usually arrive unannounced. Having a plan in place to respond immediately is critical in order to adequately document the effects of the flooding. This includes collecting perishable high water marks and obtaining remotely-sensed aerial images as quickly as possible. This data is key in verifying the accuracy of flood prediction models and equipping local officials, residents and water managers with the information they need to better identify flood risk.

ACKNOWLEDGMENTS

This project was funded through the Southwest Florida Water Management District's Watershed Management Program. We thank the pilot and crew of Aerial Cartographics of America (Orlando, FL, USA) who captured the aerial imagery of Tropical Storm Debby. We thank Amor Elder, Graduate Student Intern, at the Southwest Florida Water Management District for her work with the high-resolution imagery. The Tropical Storm Debby aerial imagery is available upon request from the Southwest Florida Water Management District, Mapping & GIS Section, 2379 Broad Street Brooksville, FL 34604, USA.

ABBREVIATIONS

The following abbreviations are used in this manuscript:

ASPRS America Society of Photogrammetry and Remote Sensing

AVHRR: Advanced Very High Resolution Radiometer

DFIRM: Digital Flood Insurance Rate Map

DEM: Digital Elevation Model

DN: Digital Number

FEMA: Federal Emergency Management Agency

fCIR: false Colour Infrared

GSD: Ground Sample Distance

H&H: Hydraulic and Hydrologic

ICPR: Interconnected Channel and Pond Routing

km/h: kilometres per hour

LiDAR: Light Detection and Ranging

mB: millibars of mercury

MODIS: Moderate Resolution Imaging Spectrometer

NAVD88: North American Vertical Datum of 1988

NIR: Near Infrared

PI: Photo-Interpreter

RGBi: Red-Green-Blue-infrared

SAR: Synthetic Aperture Radar

SPOT: Satellite Pour l'Observation del la Terra

SRTM: Shuttle Radar Topography Mission

SWFWMD: Southwest Florida Water Management District

REFERENCES

1. Sarmiento, C.; Miller, T.R. Costs and Consequences of Flooding and the Impact to the National Flood Insurance Program; American Institutes for Research: Washington, DC, USA, 2006.

2. Singhofen, P.J. Calibration and Verification of StormWaterModels. Available online: http://www.streamnologies.com/support/pdfs/Calibration.pdf (accessed on 28 December 2015).
3. Sawaya, K.E.; Olmanson, L.G.; Heinert, N.J.; Brezonik, P.L.; Bauer, M.E. Satellite Remote Sensing to Local Scales: Land and Water Resource Monitoring using High-Resolution Imagery. Remote Sens. Environ. **2003**, 88, 144–156.
4. Miller, R.L.; DelCastillo, C.E.; Chilmakuri, C.; McCorquodale, J.A.; Georgiou, I.; McKee, B.A.; D'Sa, E.J. Using multi-temporal MODIS 250 m data to calibrate and validate a sediment transport model for environmental monitoring of coastal waters. In Proceedings of the 2005 International Workshop on Analysis of Multi-Temporal Remote Sensing Images, Biloxi, MS, USA, 16–18 May 2005; pp. 2000–2004.
5. Haq, M.; Akhtar, M.; Muhammad, S.; Paras, S.; Rahmatullah, J. Techniques of Remote Sensing and GIS for Flood Monitoring and Damage Assessment: A Case Study of Sindh Province, Pakistan. Egypt. J. Remote Sens. Space Sci. **2012**, 15, 135–141.
6. Jung, H.C.; Jasinski, M.; Kim, J.W.; Shum, C.K.; Bates, P.; Neal, J.; Lee, H.; Alsdorf, D. Calibration of two-dimensional floodplain modeling in the central Atchafalaya Basin Floodway System using SAR interferometry. Water Resour. Res. **2012**, 48.
7. Tarpanelli, A.; Brocca, L.; Melone, F.; Moramarco, T. Hydraulic Modelling Calibration in Small Rivers by Using Coarse Resolution Synthetic Aperture Radar Imagery. Hydrol. Process. **2013**, 27, 1321–1330.
8. Kunyeun, H.; Youngjoo, K.; Byunhyun, K.; Famigletti, J.S.; Sanders, B.F. Calibration of Storm Water Management Model Using Flood Extent Data. Available online: https://escholarship.org/uc/item/ 88c4g4j9#page-2 (accessed on 28 December 2015).
9. Mason, D.C.; Bates, P.D.; Dall'Amico, J.T. Calibration of Uncertain Flood Inundation Models using Remotely Sensed Water Levels. J. Hydrol. **2009**, 368, 224–236.
10. Kahn, S.I.; Hong, Y.;Wang, J.; Yilmaz, K.K.; Goiurley, J.J.; Adler, R.F.; Brakenridge, G.R.; Policelli, F.; Habib, S.; Irwin, D. Satellite Remote Sensing and Hydrological Modeling for Flood Inundation Mapping in Lake Victoria Basin: Implications for Hydrologic Prediction in Ungauged Basins. IEEE Trans. Geosci. Remote Sens. **2011**, 49, 85–95.
11. Karin, F.; Petheram, C.; Marvanek, S.; Ticehurst, C.;Wallace, J.; Gouweleeuw, B. The use of hydrodynamic modelling and remote sensing to estimate floodplain inundation and flood discharge in a large tropical catchment. In Proceedings of the 19th International Congress on Modelling and Simulation, Perth, Australia, 12–16 December 2011.

10
CHAPTER

MAPPING AND MONITORING OF DUST STORMS IN IRAN BY FUZZY CLUSTERING AND REMOTE SENSING TECHNIQUES

Batool Z and Sayyad A

Keywords: Fuzzy C, Brightness temperature index, MODIS, Dust storms, Monitoring

INTRODUCTION

Dust storms are the result of turbulent winds raising large quantities of dust into the air and reducing visibility to less than 1000 m [1-3]. They are a complex process influenced by the interaction of the earth's atmosphere system, mainly caused by high wind speed, bare soil and dry air conditions [4,5].

Dust storms effect both human activity and health, and the environment, mostly due to changes in air temperatures through the absorption and scattering of solar radiation [6], which is more pronounced for an increasingly arid climate, whereby there is an observed increase in cloud condensation nuclei; a decrease of the effective radius of cloud droplets [7]. Hence, dust storms are major environmental phenomena in arid and semi-arid regions of the world, particularly in Iran.

The effects of dust storms on the world's climate system are determined and analysed by experts using numerous tools, including satellite images, weather charts, and various statistical methods. One important tool that has been used widely in recent years in order to identify, track, and analyse large-scale dust events is remote sensing [8]. The aim of this study is divided into two sections: 1) To identify the main dust storm groups in Iran by fuzzy cluster analysis and 2) To track and monitor dust storms by employing remote sensing techniques.

Citation: Batool Z, Sayyad A (2016), Mapping and Monitoring of Dust Storms in Iran by Fuzzy Clustering and Remote Sensing Techniques. J Geogr Nat Disast 6: 156, doi:10.4172/2167-0587.1000156.

Gao et al. [9] studied impacts of climate abnormality on the notable increase in dust storm events in the Hunshdak Sandy Lands, northern China between 2001–2008 and concluded that aridity is the leading factor affecting the Hunshdak dust storm frequency. Wang et al. [3] analysed the relationship between global temperatures and dust storm variation in China and observed a negative correlation between the two. Hence, global warming induces a flattening of the latitudinal thermal gradient, which subsequently results in a weakening of the surface wind strength, a possible reason for the decline of the dust storm frequency during 1954-2000. Orlovsky et al. [10] investigated dust storms in Turkmenistan and observed the highest mean annual frequency of such storms during spring within the sandy Central Karakum Desert (67 days), and a maximum number of dust storm days of 146 in western Turkmenistan. Middleton [11] mapped surface observational data to derive the frequencies of dust storms in South West Asia and determined two regions as the areas of greatest activity: an area at the borders of Iran, Afghanistan and Pakistan, comprising of the Seistan Basin, Rigistan and north-western Baluchistan; and the plains of Afghan Turkestan.

Many researchers have carried out numerous experiments to monitor dust storms via satellite remote sensing. Mei [4] monitored a dust storm event occurring in the northwest of China on April 16-17, 2007 using MODIS data. Results showed that the storm occurred within the mid-west and central areas of the Inner Mongolia province, and then moved to the east, affecting the Gansu, Shanxi, Hebei, Tianjin, Shandong provinces, with its intensity changing form weak to strong. Ochirkhuyag et al. [12] derived dust and sand storm maps from the emissive bands of MODIS and AVHRR imagery, used for the monitoring of dust and sandstorms in North – East Asia.

MATERIAL AND METHODS CASE STUDY

Iran is located between (45° E, 25° N) and (64° E, 40° N) and is surrounded by two mountain ranges, namely, the Elburz to the north and the Zagros to the west. These mountain ranges prohibit the moisture bearing systems of the Mediterranean and Caspian seas from crossing through the region in the eastern and central parts of Iran, which subsequently results in the different climatic conditions between the west and east flanks of the Zagros and between the north and south flanks of the Elburz [13]. Multiple dust sources can be identified using the annual mean map of meteorological stations data [11,14], with the Sistan Basin area found to have the most frequent dust-storm activity. The main cause of dust events in west Iran is the wind-transported dust from western neighbouring countries. In Iran, dust activity starts between April–May, peaks in June–July and greatly weakens by September. Figure 1 and Table 1 show the stations used in the study station.

Table 1: The names of studied stations and observed dust storm frequencies with visibilities less than 500m and 1000 m during the duration of this study.

Stations	Name	<1000 m	<500 m	Stations	Name	<1000 m	<500 m
1	Abadan	287	150	45	Kermanshah	44	16
2	Ahar	4	0	46	Khalkhal	0	0
3	Ahvaz	221	77	47	Khash	112	33
4	Aligudarz	3	0	48	Khoram Abad	15	1
5	Anar	22	9	49	Khoram Dareh	4	1
6	Anzali	0	0	50	Khor	23	6
7	Arak	6	2	51	Khoy	1	0
8	Ardabil	11	2	52	Kish	25	5
9	Babolsar	0	0	53	Kohrang	0	0
10	Bam	71	35	54	Maragheh	6	1
11	Bandar Lengeh	41	8	55	Mashhad	9	4
12	Bandar Mahshahr	180	73	56	Masjed Soleyman	165	44
13	Bandar Abass	86	24	57	Mianeh	1	1
14	Bijar	6	0	58	Minab	17	3
15	Birjand	12	7	59	Nehbandan	30	8
16	Bojnoord	3	2	60	Noshahr	1	0
17	Bostan	158	37	61	Pars Abad	0	0
18	Boroojen	2	1	62	Piranshahr	4	0
19	Bushehr	116	34	63	Ramhormoz	24	5
20	Chabahar	28	11	64	Ramsar	0	0
21	Dehloran	73	38	65	Rasht	0	0
22	Dezful	111	32	66	Ravansar	21	7
23	Dogonbadan	31	10	67	Sabzevar	7	2
24	Doushan Tappeh	2	0	68	Sanandaj	22	9
25	Esfahan	60	20	69	Sarab	5	3
26	Islam Abad	8	1	70	Sararud	8	2
27	Fassa	30	6	71	Saravan	67	22
28	Garmsar	4	0	72	Sarpol Zahab	39	35
29	Ghaen	2	2	73	Semnan	3	2
30	Ghazvin	22	4	74	Shahr Babak	14	5
31	Ghochan	4	0	75	Shahr Kord	4	3
32	Ghom	3	1	76	Shahrood	3	1
33	Gonabad	11	5	77	Shargh esfahan	15	8

Datasets

The main dataset used in this study is the dust-storm-frequencies with visibility less than 1000 and 500 m, defined respectively as dust storms and severe dust storms [1-3] and prevailing wind direction and speed (in knots) recorded and operated by the Iran Meteorological Organization (IRIMO) at 87 synoptic stations for the period of 1987–2013. These synoptic stations are spread throughout the country in areas that are climatically diverse (Figure 1 and Table 1). This data was classified using the statistical method of Fuzzy c-means clustering. In addition, MODIS data was used for the detection and monitoring of a severe dust storm within Iran. The MODIS sensor, on-boarded both Terra and Aqua satellites, has 36 channels with a spectral range of 0.4 μm~14.385 μm, covering the visible to the infrared portions of the electromagnetic spectrum. A total of 12 MODIS images were chosen to monitor the dust storm event that occurred in West Iran (Table 2).

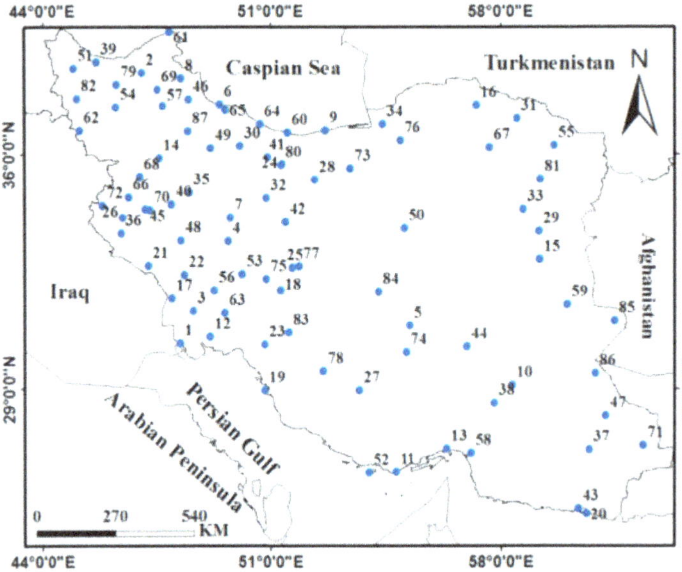

Figure 1: Spatial distribution of the selected synoptic stations within the study area.

METHODOLOGY

Fuzzy c-means algorithm

In traditional (hard partitioning) methods of clustering, feature vectors can be divided into non-overlapping clusters, with well-defined boundaries between them. Fuzzy partition clustering allows a site to belong to all regions simultaneously with a certain degree of membership. The distribution of membership of a site among the fuzzy clusters specifies the strength with which the site belongs to in each region. In recent years, advances in fuzzy set literature have spurred the development and utility of fuzzy clustering methods for a variety of applications in different disciplines. Fuzzy c-means (FCM) is a method of clustering which allows one piece of data to belong to two or more clusters to some degree that is specified by a membership grade. This method (proposed by Dunn [15] and improved by Bezdek [16]) has mostly been used in regional classification of hydrologic [17,18] and climatic variables. This kind of analysis is advantageous if there is uncertainty in determining the members of clusters. Each point has a degree of belonging to a cluster, rather than a definitive belonging to a certain cluster. Thus, points on the edge of a cluster may belong to the cluster to a lesser degree than points in the centre of the cluster [18]. The algorithm classifies a set of data

$X = X_1,\ldots,X_N$ into C groups represented as a fuzzy set $F = F_1,\ldots,F_C$. The objective of FCM is to determine the fuzzy partition of X into C clusters by minimizing the function J_m, therefore FCM performs the clustering with the aim of minimization the following objective function:

$$J_m = (U,V:X) = \sum_{i=1}^{C} \sum_{j=1}^{N} (u_{ij})^m \left\| X_j - V_i \right\|^2, (1<m<\infty)$$

where m is any real number greater than 1, u_{ij}, degree of membership of X_{jin} for the fuzzy cluster F_i, is an element of a $(C \times N)$ matrix $U = [u_{ij}]$. The ith row of U, u_i, corresponds to a fuzzy cluster F_i. $V = (V_1,\ldots,V_C)$ is a vector of cluster centroids of the fuzzy cluster $(F_1, F_2,\ldots,F_C)$. Thus, a fuzzy partition is represented by the pair (U, V). is the Euclidean norm between X_j and V_i. The parameter m controls the fuzziness of membership of each datum. FCM performs well for m within the range of 1.5 and 2.5 and thus is usually set to 2. For a given C, the sum of u_i over i must be 1. The centroid of a cluster is the mean of all points, weighted by their degree of belonging to the cluster:

$$v_i = \frac{\sum_{j=1}^{N} (u_{ij})^m x_j}{\sum_{j=1}^{N} (u_{ij})^m}$$

The degree of belonging of X_j to cluster i is proportional to the inverse distance between X_j and V_i, i.e., $d(V_i, X_j)$. The coefficients $u_{i,j}$ are normalized and fuzzed for $i=1,\ldots,C$ using:

$$u_{ij} = \left[\sum_{K=1}^{C} \left(\frac{\left\| X_j - V_i \right\|^2}{\left\| X_j - V_K \right\|^2} \right)^{1/(m-1)} \right]^{-1}$$

Fuzzy partitioning is carried through an iterative optimization of the objective function shown above, with the update of u_{ij} and cluster centres C_j such that the sum of u_j over i is 1.

Detection and monitoring of dust storms

Dust storms were extracted based on brightness temperature characteristics within the images. After atmospheric and geometric corrections, the image radiance values were converted to brightness temperature through the inverse Planck function.

$$T = \frac{hc}{\lambda_i k \ln\left(\frac{2ch^2}{L(\lambda_i)\lambda_i^5} + 1\right)}$$

where T is brightness temperature, h is Planck's constant (6.6265×10^{-34} Js), c is the velocity of light (2.998×108 m/s), k is Boltzmann's constant ($1.3806558 \times 10^{-23}$ J/K),

$L(\lambda_i)$ is the calibrated spectral radiance, and λ_i is the central wavelength (μm). Differences in dust brightness temperature at wavelengths 11 and 12 μm are used for dust extraction [19-22], which correspond to bands 31 and 32 of the imagery, respectively. The effect of dust storms in radiation reduction in band 31 is greater than that in band 32; thus, the brightness temperature for band 31 will be less than the value of band 32.

Table 2: Satellites images used in this study

Image	Date	GMT	Satellite	Longitude	Latitude	Case study
MOD021KM.A2008182.0815.005.201024103 0321	2008.06.30	08:15	Terra	27-50	18-38	18-40
MYD021KM.A2008182.0945.005.200931712 1153	2008.06.30	09:45	Aqua	44-65	May-26	37-59
MYD021KM.A2008182.0950.005.200931712 1227	2008.06.30	09:50	Aqua	37-61	23-44	
MOD021KM.A2008183.0715.005.201024123 5347	2008.07.01	07:15	Terra	43-72	33-54	18-40
MOD021KM.A2008183.0720.005.201024123 5404	2008.07.01	07:20	Terra	40-63	16-37	40-63
MOD021KM.A2008183.0855.005.201024201 1256	2008.07.01	08:55	Terra	18-44	29-50	
MYD021KM.A2008184.0935.005.200931715 2718	2008.07.02	09:35	Aqua	44-66	14-34	14-41
MOD021KM.A2008184.0805.005.201024116 4527	2008.07.02	08:05	Terra	29-50	Sep-30	31-55
MOD021KM.A2008184.0800.005.201024116 4435	2008.07.02	08:00	Terra	31-57	27-48	16-43
MOD021KM.A2008185.0705.005.201024121 4932	2008.07.03	07:05	Terra	45-70	25-46	45-61
MOD021KM.A2008185.0710.005.201024121 5123	2008.07.03	07:10	Terra	42-63	Jul-28	19-39
MOD021KM.A2008186.0750.005.201024210 3515	2008.07.04	07:50	Terra	33-56	19-39	33-56

RESULTS AND DISCUSSION

The FCM clustering was carried out on the dust storm frequencies with visibilities less than 1000 m and 500 m within the years 1987-2013, are presented in Table 1. The entire region was divided into 5 sub-regions according to two extracted indices (Figure 2). In contrast to the stations within clusters 1-4, which have confined distributions within the given regions, cluster 5 contains stations which spread out throughout the country. Hence cluster 5 is reclassified into 3 sub-groups. Figure 3 shows the reclassification of the stations for this group.

According to the clustering (Figure 4), cluster 1 includes only the individual station of Zabol, located in the southeast of Iran, northeast of the Sistan Plain and southwest to the border of Iran and Afghanistan. The Zabol station observed the maximum frequency of dust storms, reaching 670 days during study the period. Drought in the Sistan Plain within the study period reduced vegetation, led to the drying lakes of Hamoon and the desert regions of Margo, Rigistan and Khash at the borders of Iran, Afghanistan and Pakistan. In addition, winds have created suitable conditions for wind erosion and dust storms at this station. Cluster 2 includes the coastal stations of Abadan, Ahvaz (southwest Iran), Konarak (southeast Iran) and Zahedan (eastern Iran), some of which lie along the seashore. Dust storm frequencies reached up to 196-287 days in the studied period. Cluster 3 contains the stations of Bandar Mahshahr, Masjed Soleyman, Bostan, Dezful, Bushehr (the southwest Iran), Iranshahr and Khash (southeast Iran). Dust storms frequencies for this cluster were observed at 111-180. The highest frequency of dust storms in this cluster belongs to the Bandar Mahshahr station, and the lowest belongs that of Dezful. Cluster 4 includes the stations of Kermanshah, Ilam, Dehloran, Islam Abad, and Sarpol Zahab (western Iran), Yazd, Kerman, Jiroft and Bam (central Iran), Bandar Abbas, and Bandar Lengeh (southern Iran), Dogonbadan (southwest Iran), Nehbandan (eastern Iran), Saravan and Chabahar (southeast Iran). This cluster had an observed 26-86 days of dust storms. The lowest and highest dust storm frequencies observed in Chabahar and the Bandar Abbas stations respectively. In total, clusters 1-4 represent the stations that have a concentrated spread from the northwest to the southeast, lying east of the flank of the Zagros mountains range, separated due to their proximity to the important dust source regions to the east in Rub al Khali and the Arabian deserts. Clusters 5-1, 5-2 and 5-3 include the northern and southern slopes of the Elburz Mountain range; the north-western country, the high elevations of the Zagros Mountain range and the northeast of Iran. The frequency of dust storms in these clusters are 0-24 days. Stations located in each of these clusters are not homogeneous in terms of geographical position and climate and dust source but are homogeneous in severity and frequency of dust storms. Generally, the frequency of dust storms decreases from lowlands and lower elevations of the west towards high elevations and the eastern slope of Zagros, the northwest and southwest and decrease from the east and southeast toward central, south and Northeast of Iran (Figure 4). This is directly related to the distance of these stations from the sources of the dust, their location in the path of the external synoptic systems, physical and geomorphologic features and local climate.

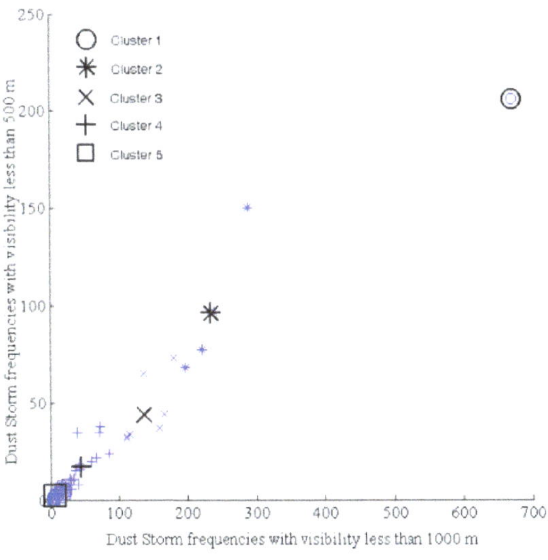

Figure 2: Clustered data set using the fuzzy c-means algorithm.

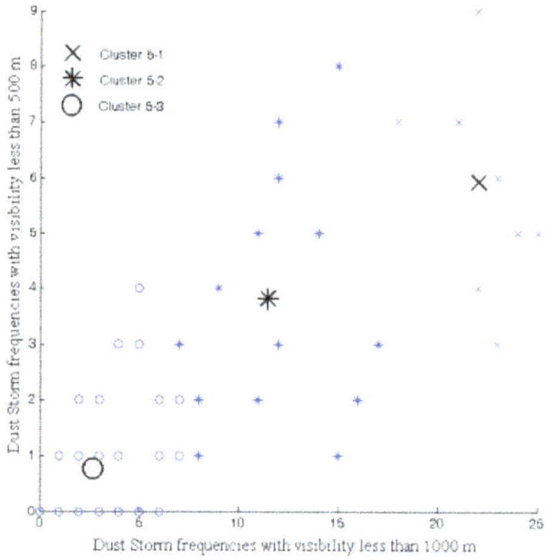

Figure 3: Re-clustering of cluster 5 using the fuzzy c-means algorithm.

One of the most severe dust storms occurred from 30 June 2008 to 4 July 2008. This dust storm was selected and analysed to track and monitor the dust movement. Over 12 provinces were affected and the visibility declined to less than 500 m at the most affected stations. Extraction of the dust storm aimed distinguish dust storm from cloud, snow, ground, etc. and to accurately depict dust storm boundaries. It is known that the brightness temperature (T) of dust regions is more than 1 kelvin, with other objects less than 1k. Hence, the threshold of 1k was to separate the dust from other objects.

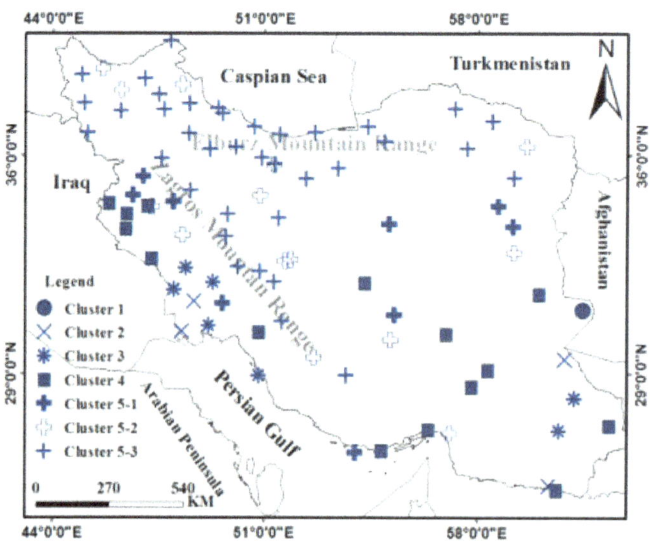

Figure 4: Location of fuzzy clusters obtained using the fuzzy clustering analysis.

12 MODIS images were chosen to monitor the scope and movement of the dust storm. Result show that the dust storm occurred in Iraq and south of Arabian Peninsula on 30 June 2008, and then moved to the east on 1 July, affecting the west of Iran, in particular the provinces Khuzestan, Bushehr, Ilam, Kohgeluye va Boyir Ahmad. Within the following days, dust storm had spread on the central and south Iran (Figure 5). Provinces that were severely affected by a severe storm were mainly located in the western station clusters of 2, 3 and 4.

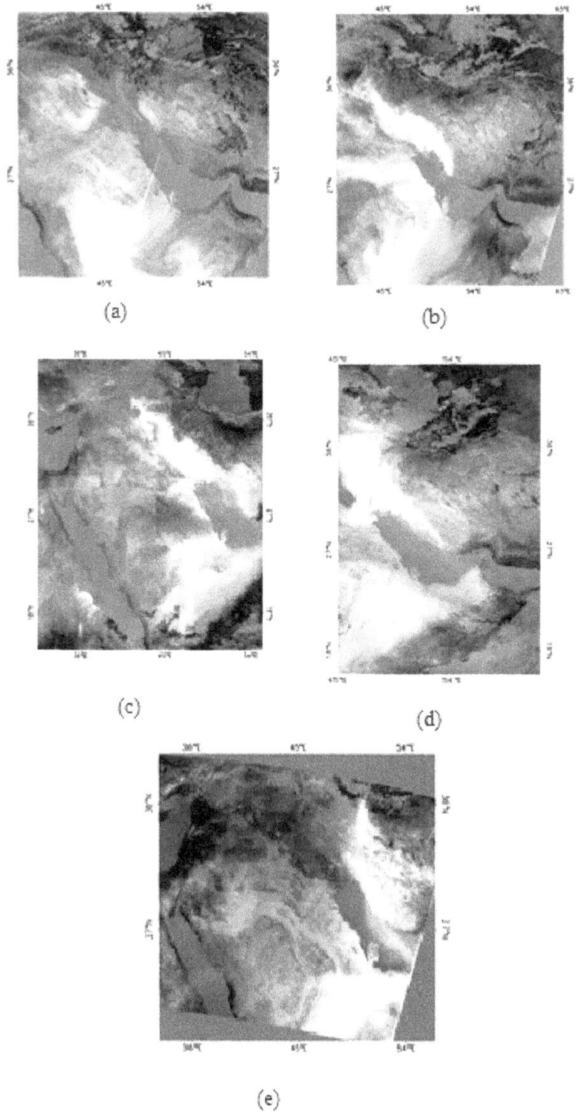

Figure 5: Result of monitoring the dust storm on days of 30 Jun (a), 1 Jul (b), 2 Jul (c), 3 Jul (d), 4 Jul (e).

Interpretation of the remotely sensed data confirmed the results of the fuzzy classification of the stations based on dust storm frequencies. The results of dust storm tracing via the remote sensing technique indicated that the west and southwest parts of the country are strongly influenced by atmospheric circulation patterns generating form the Sahara, Rub al Khali and Arabian deserts. This is confirmed by the prevailing wind direction of studied stations in the western half of Iran (Table 3). In the eastern half of Iran strong seasonal and local winds are observed, mainly in spring and summer, due to local pressure gradients with prevailing wind that pass from the east, north east, central deserts and also the drying bed of Hamoon lakes Iran, creating dust storms (Table 3). Dodangeh et al. [18] also showed that variability of dust storm frequencies within this region shows different pattern from that of other parts of the country. They showed that regional growth curves (RGCs) for west sides of the Zagros mountain ranges presented a sharper slope compared to the east sides. We also demonstrated that visibility indices are different for the stations included in this region and remaining stations of the country. By visual inspection of Figure 5, is revealed that dust storms generated in the Arabian Peninsula extend toward the country and affect west and northwest regions [23]. Separated grouping of the stations in this region is mainly attributed to this process (Table 3).

CONCLUSION

Iran is affected by dust storms repeatedly due to its location in an arid and semi-arid belt of the world. Therefore, the identification and clustering of this phenomenon is essential and a priority for climate studies. According to the results, Iran can be classified into five clusters by the dust-storm-frequencies, from which cluster 5 is reclassified in three clusters due to its wide range. Maximum frequencies of storms were observed at the stations of Zabol (cluster 1), Abadan, Konarak, Ahwaz, Zahedan (cluster 2), Dezful, Bandar Mahshahr, Bostan, Masjed Soleyman, Bushehr, Khash, Iranshahr (cluster 3). Minimum cores dust storms (cluster 5-1) include stations in areas of north, northwest, northeast Iran and also the higher elevations of the Zagros in western Iran, representing 0-7 days with dust storms. In general, dust storm frequencies decrease from the lowlands and low elevations of the west and south west toward Zagros high elevations and east, north, northwest, and south Iran, as well as from east and southeast toward centre, south and northeast Iran. A case study of a severe dust storm event using satellite images indicates that the Brightness Temperature Index is a suitable index for the detection and monitoring of dust storms. Sources of dust storms and their tracks can be identified by this method. The source of investigated dust storm was found to be in Iraq and south of the Arabian Peninsula, which considerably influence the western half of Iran, containing stations from clusters 2, 3 and 4.

Table 3: Prevailing wind direction and speed (In Knots) at stations during the study period

Stations	Name	Prevailing Wind Direction	Prevailing Wind Speed (In Knots)	Stations	Name	Prevailing Wind Direction	Prevailing Wind Speed (In Knots)
1	Abadan	315° (NW)	10.1	45	Kermanshah	270° (W)	8.1
2	Ahar	270° (W)	10.6	46	Khalkhal	90° (E)	10.9
3	Ahvaz	270° (W)	7.1	47	Khash	270° (W)	10.9
4	Aligudarz	135° (SE)	10.3	48	Khoram Abad	270° (W)	10.9
5	Anar	225° (SW)	10.6	49	Khoram Dareh	315° (NW)	10.9
6	Anzali	360° (N)	7.2	50	Khor	45° (NE)	10.9
7	Arak	270° (W)	9.2	51	Khoy	45° (NE)	6.9
8	Ardabil	90° (E)	13.7	52	Kish	270° (W)	10.8
9	Babolsar	270° (W)	6	53	Kohrang	270° (W)	9.1
10	Bam	270° (W)	8.5	54	Maragheh	90° (E)	10
11	Bandar Lengeh	90° (E)	9.7	55	Mashhad	180° (S)	10.9
12	Bandar Mahshahr	315° (NW)	11.5	56	Masjed Soleyman	270° (W)	9.5
13	Bandar Abass	180° (S)	9.1	57	Mianeh	45° (NE)	9
14	Bijar	180° (S)	11.6	58	Minab	225° (SW)	10.9
15	Birjand	90° (E)	7.2	59	Nehbandan	315° (NW)	17
16	Bojnoord	270° (W)	12.2	60	Noshahr	270° (W)	10.9
17	Bostan	270° (W)	9.4	61	Pars Abad	90° (E)	9.5
18	Boroojen	270° (W)	9.4	62	Piranshahr	225° (SW)	10.8
19	Bushehr	315° (NW)	9	63	Ramhormoz	315° (NW)	10.9
20	Chabahar	135° (SE)	8.7	64	Ramsar	315° (NW)	10.9
21	Dehloran	270° (W)	9.7	65	Rasht	270° (W)	10.9
22	Dezful	225° (SW)	8	66	Ravansar	315° (NW)	9.9
23	Dogonbadan	270° (W)	9.4	67	Sabzevar	90° (E)	10.9
24	Doushan Tappeh	45° (NE)	6.1	68	Sanandaj	180° (S)	10.9
25	Esfahan	270° (W)	7.3	69	Sarab	90° (E)	9.7
26	Islam Abad	270° (W)	8	70	Sararud	270° (W)	10.7
27	Fassa	315° (NW)	6.6	71	Saravan	45° (NE)	10.9
28	Garmsar	90° (E)	8.6	72	Sarpol Zahab	270° (W)	7.8
29	Ghaen	45° (NE)	9.5	73	Semnan	360° (N)	10.9
30	Ghazvin	135° (SE)	8.1	74	Shahr Babak	225° (SW)	10.9
31	Ghochan	135° (SE)	6.8	75	Shahr Kord	225° (SW)	9
32	Ghom	270° (W)	8	76	Shahrood	45° (NE)	10.9
33	Gonabad	90° (E)	8.1	77	Shargh esfahan	270° (W)	8.2
34	Gorgan	270° (W)	7.1	78	Shiraz	315° (NW)	10.9
35	Hamadan	225° (SW)	10.4	79	Tabriz	90° (E)	8.6
36	Ilam	270° (W)	8.1	80	Tehran	270° (W)	10.9
37	Iranshahr	270° (W)	8.7	81	Torbate Heydarieh	90° (E)	10.9
38	Jiroft	180° (S)	10.9	82	Urmia	270° (W)	5.4
39	Jolfa	45° (NE)	10.7	83	Yasuj	180° (S)	10.9
40	Kangavar	225° (SW)	11.4	84	Yazd	315° (NW)	10.9
41	Karaj	315° (NW)	10.9	85	Zabol	315° (NW)	17
42	Kashan	45° (NE)	10.9	86	Zahedan	360° (N)	9.6
43	Konarak	225° (SW)	10.9	87	Zanjan	90° (E)	10.9
44	Kerman	360° (N)	10.9	-	-		

REFERENCES

1. 1. McTainsh GH, Pitblado JR (1987) Dust storm and related phenomena measured from meteorological record in Australia. Earth Surf Process Landforms 12: 415-424.
2. 2. Goudie AS, Middleton NJ (2006) Desert Dust in the Global System. Springer pp: 1-287.
3. 3. Wang LA, Shichang K, Pang Deqian (2009) On The Relationship between Global Warming and Dust Storm Variation in China. International Conference on Environmental Science and Information Application Technology pp: 59-62.
4. 4. Mei D, Xiushan L, Lin S, Ping W (2008) A Dust-Storm Process Dynamic Monitoring With Multi-Temporal MODIS Data. The International Archives of the Photogrammetry, Remote Sensing and Spatial Information Sciences 37: 965-970.
5. 5. Xingkui X, Levy JK, Zhaohui L, Hong C (2006) An investigation of sand –dust storm events and land surface characteristics in China using NOAA NDVI data. Global and Planetary Change 52: 182-196.
6. 6. Goudie AS (2009) Dust storms: Recent developments. Journal of Environmental Management 90: 89-94.
7. 7. Han Y, Dai X, Fang X, Chen Y, Kang F (2008) Dust aerosols: a possible accelerant for an increasingly arid climate in North China. J Arid Environ 72: 1476-1489.
8. 8. Azizi GH, Shamsi pour AA, Miri M, Safarrad T (2012) Synoptic and remote sensing analysis of dust events in southwestern Iran. Nat Hazards 64:1625-1638.
9. 9. Gao T, Han J, Wang Y, Pei H, Lu S (2011) Impacts of climate abnormality on remarkable dust storm increase of the Hunshdak Sandy Lands in northern China during 2001–2008. Meteorological Applications pp: 265-278.
10. 10. Orlovsky L, Orlovsky N, Durdyev A (2005) Dust storms in Turkmenistan. Journal of Arid Environments 60: 83-97.
11. 11. Middleton NJ (1986) A Geography of dust storms in southwest Asia. Journal of Climatology 6: 183-96.
12. 12. Ochirkhuyag L, Tsolmon R (2008) Monitoring the Source of Trans-National Dust Storms in North East Asia. The International Archives of the Photogrammetric, Remote Sensing and Spatial Information Sciences 37: 835-840.
13. 13. Dodangeh E, Shao Y, Daghestani M (2012) L-Moments and fuzzy cluster analysis of dust storm frequencies in Iran. Aeolian Research 5: 91-99.
14. 14. Washington R, Todd M, Middleton N, Goudie AS (2003) Global dust storm source areas determined by the Total Ozone Monitoring Spectrometer and ground observations. Ann Assoc Am Geogr 93: 297-313.
15. 15. Dunn JC (1974) A Fuzzy relative of the ISODATA process and its use in detecting compact, well-separated clusters. J Cybernetics 3: 32-57.

16. 16. Bezdek JC (1981) Pattern recognition with fuzzy objective function algorithms. Plenum Press, New York.
17. 17. Rao AR, Srinivas VV (2006) Regionalization of watersheds by fuzzy cluster analysis. Journal of Hydrology 318: 57-79.
18. 18. Dodangeh S, Sattari M, Seckdn N (2011) Regional frequency analysis of minimum flow by L-Moments method. Journal of Agricultural Science 17: 43-58.
19. 19. Zhang P, Lu NM, Hu XQ, Dong CQ (2006) Identification and physical retrieval of dust storm using three MODIS thermal IR channels. Global and Planetary Change 52: 197-206.
20. 20. Baddock MC, Bullard JE, Bryant RE (2009) Dust source identification using MODIS: A comparison of techniques applied to the Lake Eyre Basin. Australia Remote Sensing of Environment 113: 1511-1528.
21. 21. Ackerman SA (1997) Remote sensing aerosols using satellite infrared observations. Journal of Geophysical Research 102: 17069-17079.
22. 22. Sokolik I (2002) The spectral radiative signature of wind-blown mineral dust: Implications for remote sensing in the thermal IR region. Geophysical Research Letters 29: 2154.
23. 23. Qu JJ, Kafatos M (2006) Asian dust storm monitoring combining Terra and Aqua MODIS SRB measurements. Geosciences and Remote Sensing letters 3: 484-486.

11

CHAPTER

DETECTION OF SHORELINE CHANGES ON THE VISAKHAPATNAM COAST, ANDHRA PRADESH, FROM MULTI-TEMPORAL SATELLITE IMAGES

Kannan R, Kanungo A and Murthy MVR

Keywords: Shoreline changes, Hudhud, Satellite images, DSAS; Coastal processes

INTRODUCTION

Coastlines are dynamic morphological entities that respond to the external forces exerted by waves, tides, nearshore currents and resultant sediment transport. When the resultant sediment transport entering a particular area is greater than the sediment going out from the area, accretion or beach development takes place [1]. Conversely, when there is a deficit of the incoming sediment supply, beach erosion takes place [2]. Beaches act as constant absorbers of the wave energy of water and though subject to small disturbances, remain in equilibrium. However, sometimes this equilibrium gets disturbed due to either natural phenomena or human intervention [3]. When shore structures are constructed, it is quite likely that the equilibrium is destroyed. Since this can cause considerable damage and reduce the effectiveness of such structures, it is necessary to study the equilibrium conditions of shores before constructing such structures [4].

Occasionally, coastal erosion processes can be very expansive and devastating to invaluable properties, human lives and even the natural environment. Globally, this has generated much concern and interests are on the increase in academic discourse [5]. The natural action of winds and waves, together with the anthropogenic forces resulting from the continued desire for natural resource exploitation, are constantly at work in this region. Although human actions may sometimes yield positive results,

Citation: Kannan R, Kanungo A, Murthy MVR (2016) Detection of Shoreline Changes Visakhapatnam Coast, Andhra Pradesh from Multi-Temporal Satellite Images. J Remote Sensing & GIS 5:157, doi:10.4172/2469-4134.1000157

they cannot be completely exempted from facilitating and accelerating the extent of damage to the natural landscape [6]. For coastal zone monitoring, shoreline extraction is fundamental. The shoreline, which is defined as the position of the land-water interface at one instant in time [7], is a highly dynamic feature and is an indicator for coastal erosion and accretion. The processes of erosion and accretion affect human life, cultivation and natural resources along the coast [8]. Rapid shoreline changes can create catastrophic social and economic problems along populated strands. Design of viable land-use and protection strategies to reduce potential loss is necessary, and this requires comprehension of regional shoreline dynamics [9].

Since coastal areas are regions of high economic value, the prediction of shoreline positions depends solely on having a clear understanding of the shoreline parameters [10]. Based on this argument therefore, an appreciable knowledge of the shoreline characteristics is of utmost importance. This has become essential to make informed decisions towards effective coastal management [11]. If such parameters are put in place, it is believed that any information relative to shoreline characteristics will be readily accessible at any point in time. In the light of the foregoing, taking into consideration the high economic potentials of the area, this study seeks to extract shorelines from the satellite imagery, determine the rate of shoreline change, as well as the net shoreline movement in the area [12].

STUDY AREA

The Visakhapatnam district is the 7th largest district in Andhra Pradesh, which accommodates the second largest city of Andhra Pradesh, occupying an area of approximately 11,161 km^2. As Visakhapatnam is an important port city in the eastern coast of India, it attracts many tertiary sectors. The district is the central hub for industry and education in the state of Andhra Pradesh, and is home to the Eastern Naval Command. Apart from Visakhapatnam, Gangavaram Port also serves as a gateway for contributing to the development of the petroleum, steel and fertilizer industries in the district. It is a hub for iron ore and other mineral exports that are exported through sea route to other countries (Figure 1).

The district is situated between 81°06′ and 83°31'E and 17°15′ and 18°32'N. The boundaries of this district are Bay of Bengal in the east, the East Godavari district in the south and the Orissa state in the west and north directions. The length of the coastline is 135 km. Headland bay and other small beaches are found in the Vishakhapatnam district. Fishing is a major sector on which the economy of the district, as well as the state of Andhra Pradesh, depends on. The district has one of the largest harbours in the country for the fishing industry and seafood exports. A number of fishermen have their livelihoods depending on fishing in the district. The ice factories in and outside the fishing harbour, caters to the needs of the fishermen, as well as providing employment.

MATERIAL AND METHODOLOGY

This study is carried out using Multi-Resolution and Multi-temporal satellite data from Landsat TM and IRS. These datasets were acquired on cloud free days over different dates during the study period (1989 to 2015). The multi-temporal capabilities allow tracking of changes over time. Survey of India (SOI) topographic sheets (1:50,000 scale) were used for preparing the base map (Table 1).

Shoreline extraction and delineation

Recent advancements in remote sensing and geographical information system (GIS) techniques have led to improvements in coastal geomorphological studies, such as automatic and semi-automatic methods [13]. The semi-automatic techniques are attractive due to their cost-effectiveness, minimal input times etc. Automatic shoreline delineation is a complex process due to the presence of water saturated zones at the land-water boundary [14]. Shorelines were identified and delineated by processing the NIR bands of Landsat TM using "Grey Level Thresholding" and the "Edge Enhancement Technique". In this present study, the exact land-water boundary was obtained by using a "non-linear edge-enhancement" technique; these operations were applied to the image data to produce an enhanced image output for subsequent visual interpretations [6]. The enhancement techniques improve the feature exhibition and increases visual distinctions between features contained in a scene.

To extract the shorelines from the satellite images, shape files were created in ArcCatalog for each of the images [15]. For easy data handling, the images were spatially projected onto the UTM coordinate reference system. This was followed by the determination of shoreline reference features. The high-water line (HWL) was therefore used as it was relatively easy to distinguish on all the images as a wet/dry line.

Sensor	Spatial Resolution	Year	Month	Source
Landsat 5TM	28.5	1989	November	USGS
IRS-LISS III	23.5	1999	January	NRSC
IRS-P6 LISS III	23.5	2005	February	NRSC
IRS-P6 LISS III	23.5	2010	January	NRSC
IRS-P6 LISS IV	5.8	2012	March	NRSC
Landsat 8TM	28.5	2015	March	USGS

Table 1: The datasets used in this study. The images were processed to delineate the shorelines for the years of 1989, 1999, 2005, 2010, 2012 and 2015, in order to determining their rate of changes over the study period.

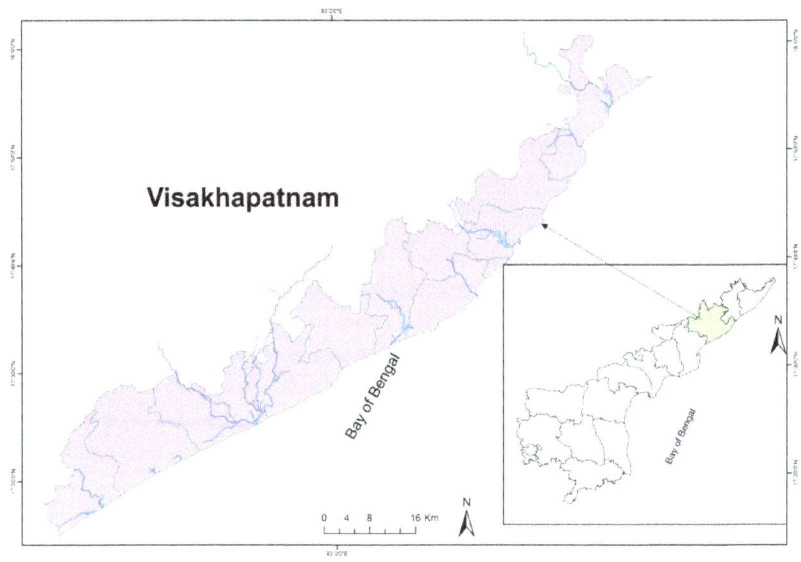

Figure 1: Study area.

The extraction was then carried out using the heads-up digitizing method. This manual method was adopted in an attempt to avoid the difficulties associated with the use of automated methods of extraction [16]. Features from the satellite images were digitized along the dry-wet sand boundaries, which could be recognized from different tones on the sand beach. Usually, the tonal differences are caused by the variation in moisture in the sands, as a result of being previously immersed or washed by high water levels (Figure 2).

Digital shoreline analysis system (DSAS)

The technique calculates the rates of changes, such as erosion or accretion, along shorelines. The calculation is based on the ArcGIS DSAS extension, after the extraction of the coastline from the Landsat TM and IRS images [17]. The DSAS is a computer software that is able to compute the rate-of-change statistics from multiple historic shoreline positions. DSAS generates transects that are cast perpendicular to the baseline along-shore at a user-specified spacing [18]. The transect shoreline intersections along this baseline are then used to calculate the rate-of-change statistics. Based on the DSAS settings, 676 transects were generated, perpendicular to the baseline, at a 200 m spacing and approximately 135 km along the Visakhapatnam shoreline. DSAS 4.2 has 6 statistical approaches to compute changes. In this study, the End Point Rate (EPR) and the Linear Regression Rate (LRR) approaches were used. LRR uses all the available data to compute long-term rates of change. The EPR is calculated by dividing the distance of shoreline movement by the time elapsed between the oldest and the most recent shoreline. The LRR statistic can be determined by fitting a least-squares regression line to all shoreline points for a particular transect [19].

RESULTS AND DISCUSSION

The shoreline change estimation was carried out for entire Visakhapatnam coast, at approximately 135 km in length. The shorelines were extracted from multi-temporal satellite images for 1989, 1999, 2005, 2010, 2012 and 2015 by via digital image processing and visually interpretation. The coastline rate of change was calculated using the DSAS model with two different statistical techniques; End Point Rate (EPR) and Linear Regression Rate-of-Change (LLR). Base lines were constructed seaward of and parallel to the general trend of the shorelines. The transects were spaced 200 m apart and the rates of shoreline change were calculated at each transect using linear regression. Results show that the most significant changes were observed at the Visakhapatnam Coast (Table 2).

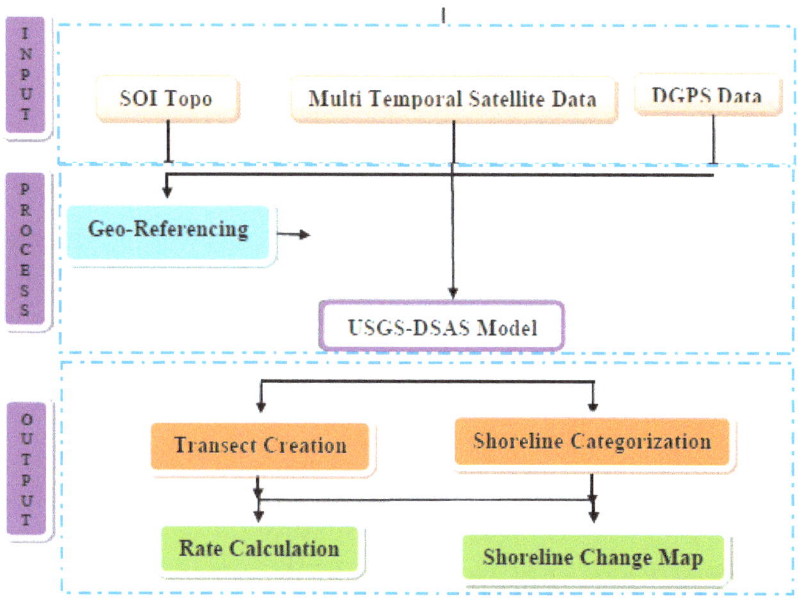

Figure 2: Methodology.

Shoreline status (Rate of change)

For the period 1989 to 1999, shoreline maximum locations had eroded, due a to severe cyclone event in Visakhapatnam coast in the year of 1998. The maximum erosion was observed at -9.7 m/yr, north of the Gajuwaka area, and maximum accretion at +17.3 m/yr, south of the Gangavaram port. The average erosion was estimated as -2.6 m/yr and accretion as +2.1 m/yr.

A combination of accretion and erosion was observed for the period of 1999 to 2005. During this time, the south side of Appikonda village up to the Annavaram coast had been fully eroded, while south side of Gangavaram port developed a fully accretion coast.

Classification of the Coast	Extend (Km)	Percent of the Coast	Cumulative (%)
Length of the coast including River and Port	135.2		
High Erosion	0.2	0.15	
Medium Erosion	9	6.66	
Low Erosion	22	16.27	23.08
Artificial coast: Seawall, Riprap	5.2	3.85	26.92
Stable Coast	24.2	17.90	17.90
Low Accretion	69.2	51.18	
Medium Accretion	5	3.70	
High Accretion	0.4	0.30	55.18
No. of port/Harbour	2		
No. of Groins	5		

Table 2: Shoreline classification Status.

The maximum erosion was found to be at -20.62 m/yr and the accretion at 11.76 m/yr. The average erosion and accretion was estimated at -2.97 m/yr and 3.30 m/yr respectively, with an overall -136 m eroded and 70.93 m accretion (Figure 3).

During the years of 2005 to 2010, the coast had eroded due to the Tsunami of December 2004. The maximum erosion was observed as -20.2 m/ yr, in the Annavaram areas, and maximum accretion as +10.2 m/yr, in the Payakaraopeta village. The average erosion for this period was calculated as -3.4 m/yr and accretion as +2.2 m/yr. The net shoreline movement was given as -99.96 m/yr eroded and 50.18 m accreted. Accretion was predominant for the 2010-2012 period. The maximum erosion was observed as -29.4 m/yr in the Rambilli area and accretion as +46.0 m/yr, in the north side of the Visakhapatnam port area. The average erosion was observed as -6.6 m/yr and accretion as 14.3 m/yr. The net shoreline movement for this period was a 100.77 m accretion and -64.37 m erosion (Figure 4).

During 2012-2015, mostly accretion was observed, however there was some erosion due to the HUDHUD cyclone in October 2014. The maximum erosion was estimated as -22.39 m/yr in the north side of Bheemunipatnam, with a maximum accretion of 30.25 m/yr in the south side of Gangavaram and the north side of Visakhapatnam port areas. The overall erosion and accretion for above period was -5.35 m/yr and 8.37 m/ yr, respectively, with the movement of net shoreline erosion and accretion estimated as -66 m and 89 m (Figure 5).

Shoreline change using linear regression rate (LRR) method

A linear regression rate-of-change statistic can be determined by fitting a least-squares regression line to all shoreline points for a particular transect [17]. The regression line is placed so that the sum of the squared residuals is minimized. The linear regression rate is the slope of this line. The method of linear regression includes the following features: (1) all the data is used, regardless of changes in trend or accuracy, (2) the method is purely computational, (3) the calculation is based on accepted statistical concepts, and (4) the method is easy to employ [20]. However, the linear regression method is susceptible to outlier effects and also tends to underestimate the rate of change relative to other statistics, such as EPR [17,21].

The Linear Regression Rata (LRR) of shoreline change were calculate at each transect. For this study, a total of 576 transects were determined, with a spacing of 200 m. The highest erosion rate, at -5.02 m/yr, was found to be in Bheemunipatnam coast and highest accretion of +10.31 m/yr in south side of the Gangavaram port area [22]. The net shoreline movement was observed as 95 m of erosion, and 147 m of accretion (Figure 6).

Shoreline change statistics along Visakhapatnam coast

The total length of the Visakhapatnam coast is approximately 135 km. Between years of 1989 to 2015, accretion and erosion was identified at 74.6 km and 31.2 km respectively, and a total of 24.6 km as stable. Figure 7 and Table 3 show detailed shoreline change statics for each period wise.

SUMMARY AND CONCLUSIONS

The Visakhapatnam coast is one of the most important public beaches along the Andhra Pradesh coast. In a natural environment, erosion and accretion at a certain area are normally in balance. If erosion occurs at a particular area, accretion should also be found. In this study, this is not the case, proving that the sediment transport in the area has been disturbed. This study clearly demonstrates that the integration of remote sensing and GIS technology is very useful for long term shoreline change studies with reasonable accuracy. The result of shoreline change mapping will be valuable for coastal engineers and coastal zone management authorities to facilitate suitable management plans and the regulation of coastal zones.

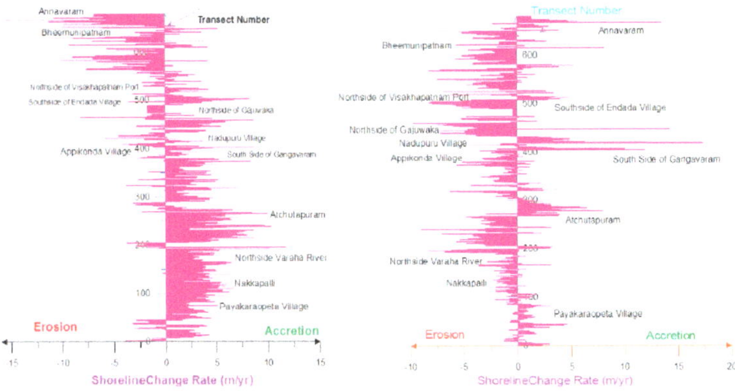

Figure 3: Shoreline change status 1989-1999 and 1999-2005.

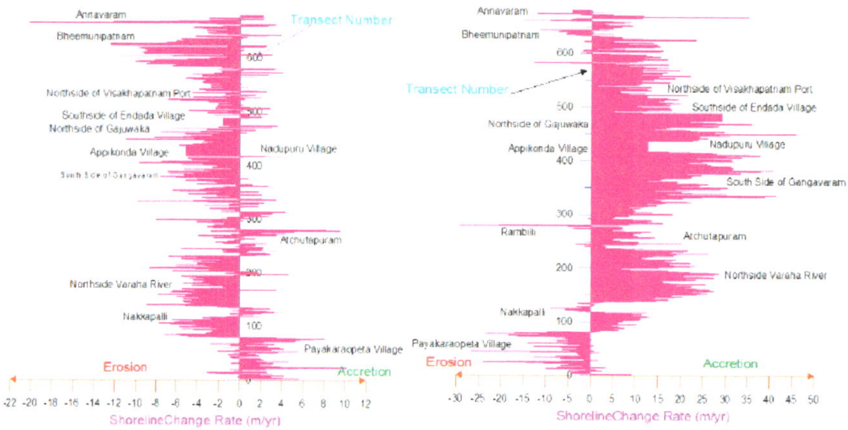

Figure 4: Shoreline change status 2005-2010 and 2010-2012.

Figure 5: Shoreline change status 2012-2015.

Figure 6: Shoreline change status using the LRR method (1989-2015).

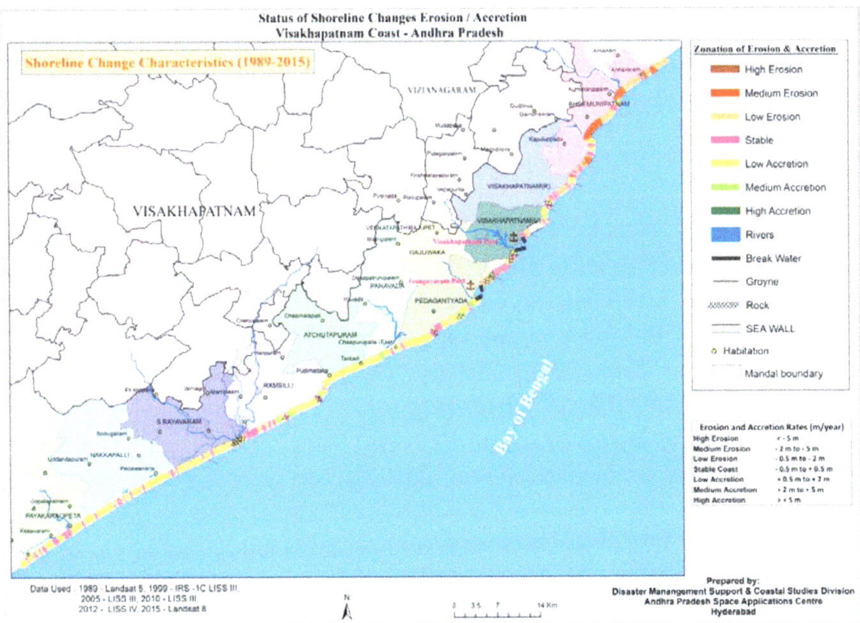

Figure 7: Shoreline classification along the Andhra Pradesh coast.

Shoreline Classification	Shoreline Length (1989-99) Km	Shoreline Length (1999-2005) Km	Shoreline Length (2005-10) Km	Shoreline Length (2010-12) Km	Shoreline Length (2012-15) Km
High Erosion	11	8	21	10.8	14.8
Medium Erosion	32.6	13.4	38.6	7.4	9
Low Erosion	29.2	14.4	23	4.2	7
Stable Coast	21.4	11.8	12.6	1.6	5.8
Low Accretion	19	22	13.2	4.4	9.2
Medium Accretion	14.2	44	16	9.6	21.6
High Accretion	3.2	16.6	2.8	89.4	60

Table 3: Classification along the Andhra Pradesh coast.

ACKNOWLEDGEMENTS

I would like to thank Mr. Sanjay Gupta, (I.F.S.), and Director General of Andhra Pradesh Space Applications Centre for providing this opportunity to carry this research. The authors also like to thank APSAC Disaster management support and coastal studies Division (DMS&CS) for encouragement for this work.

REFERENCES

1. 1. Baily B, Nowell D (1996) Techniques for monitoring coastal change: a review and case study. Ocean and Coastal Management 2: 85-95.
2. 2. Mani Murali R, Shrivastava D, Vethamony P (2009) Monitoring shoreline environment of Paradip, East coast of India using remote sensing. Current Science 97: 79-84.
3. 3. Douglas BC, Crowell M (2000) Long-term shoreline position prediction and error propagation. J of Coastal Research pp: 145-152.
4. 4. Alesheikh AA, Ghorbanali A, Nouri N (2007) Coastline change detection using remote sensing. International Journal of Environment Science and Technology pp: 61-66.
5. 5. Anthony EJ (2005) Beach Erosion: Encyclopedia of Coastal Science. Springer pp: 140-144. 6. Crowell M, Leatherman SP (1999) Coastal erosion mapping and management. Journal of Coastal Research 28: 1-196.
6. 7. Gopinath G, Seralathan P (2005) Rapid erosion of the Coast of the Sagar Island, West Bengal. Environmental Geology 48: 1058-1067.
7. 8. Fletcher C, Rooney J, Barbee M, Lim SC (2003) Mapping Shoreline Change Using Digital Orthophotogrammetry on Maui, Hawaii. Journal of Coastal Research pp: 106-124.
8. 9. Blodget HW, Taylor PT, Raork JH (1991) Shoreline changes along the rosetta nile promontory - monitoring with satellite-observations. Marine Geology 99: 67-77.
9. 10. Kaliraj S, Chandrasekar N, Magesh NS (2013) Impacts of wave energy and littoral currents on shoreline erosion/accretion along the south-west coast of Kanyakumari. Tamil Nadu using DSAS and geospatial technology. Environ Earth Sci 71: 4523-4542.

10. 11. Pandian PK, Ramesh S, Murthy MVR, Ramachandran S, Thayumanavan S (2004) Shoreline changes and near-shore processes along Ennore Coast, East Coast of South India. Journal of Coastal Research 20: 828-845

11. 12. Nayak SR (2002) Use of satellite data in coastal mapping. Indian Cartographer 22: 147-157.

12. 13. Moore LJ (2000) Shoreline mapping techniques. Journal of Coastal Research 16: 111-124.

13. 14. Stauble DK (2003) The use of shoreline change mapping in coastal engineering project assessment. Journal of Coastal Research pp: 178-206.

14. 15. Maiti S, Bhattacharya AK (2009) Shoreline change analysis and its application to prediction: A remote sensing and statistics based approach. Marine Geology 257: 11-23.

15. 16. ICMAM (2005) Shoreline Management Plan for Ennore Coast (Tamilnadu). Report of Integrated Coastal and Marine Area Management Project Directorate, Ministry of Earth Sciences, India.

16. 17. Thieler ER, Himmelstoss EA, Zichichi JL, Ergul A (2009) The Digital Shoreline Analysis System (DSAS) version 4.0 an ArcGIS extension for calculating shoreline change, U.S. Geological Survey.

17. 18. Saranathan E, Chandrasekaran R, Manickaraj SD, Kannan M (2011) Shoreline Changes in Tharangampadi Village, Nagapattinam District, Tamil Nadu, India-A Case Study. Journal of Indian Society Remote Sensing 39: 107-115.

18. 19. Rao PP, Nair M, Raju DV (1985) Assessment of the role of Remote Sensing in monitoring shoreline changes: a case study of the Kerala Coast. International Journal of Remote Sensing 6: 549-558.

19. 20. Dolan R, Hayden BP, May S (1991) Erosion of the US shorelines. Journal of Remote Sensing 6: 26-31.

20. 21. Dolan R, Hayden B, May P, May S (1980) The reliability of shoreline change measurements from aerial photographs. Shore and Beach pp: 22-29.

21. 22. Kannan R, Anand KV, Sundar V, Sannasiraj SA, Rangarao V (2014) Shoreline changes along the Northern coast of Chennai port, from field measurements. ISH Journal of Hydraulic Engineering 20: 24-31.

12
CHAPTER

IMPACTS OF RAINSTORMS DURING AUSTRAL WINTER IN SAO PAULO STATE, BRAZIL: A CASE STUDY

Franchito SH, Gan MA, Rao VB, Santo CME, Conforte JC and Pinto O

Keywords: Sao Paulo city, Rainstorms; Damages

INTRODUCTION

Southeast Brazil (SEB) is characterized by strong annual variation of rainfall, with rainstorms in austral summer and light rainfall in austral winter. This annual variation in rainfall is linked to the monsoon regime over South America [1-4]. Episodes of rainstorms with strong winds in this region may have drastic economic and social impacts. SEB is the most populous region and plays an important role in the economy of the country, being characterized by high industrial activity, agricultural productivity and hydroelectric generation.

Some studies have investigated the occurrence of severe weather in SEB [5-8]. Although rainstorms and strong winds are concentrated mainly in austral summer, they could also be present during austral winter, the dry season. Very few references in the literature can be found on rainstorms in SEB during the austral winter months. Seluchi et al. [7] studied the mechanisms of rainstorms that occurred between 19 and 20 July 2004 in Serra do Mar, located in SEB along the coastal region. This event represented one of the most extreme landslide cases in the past 15 years in the region. It was found that the persistence and configuration of the low–level circulation played an important role in the middle troposphere dynamics and the thermodynamic instability. Sias et al. [9] studied a convective event which occurred during winter 1989 in São Paulo city. It was observed that winter precipitation over 50 mm day^{-1} occurred, on average, once in every ten years and was mainly related to the passage of cold fronts.

Citation: Franchito SH, Gan MA, Rao VB, Santo CME, Conforte JC, et al. (2016),s Impacts of Rainstorms during Austral Winter in Sao Paulo State, Brazil: A Case Study. J Geogr Nat Disast 6: 162, doi:10.4172/2167-0587.1000162.

Although the austral winter months from June to August 2015 were very dry, during September 2015 many episodes of rainstorms and strong winds occurred in SEB, particularly over Sao Paulo state (53°W-44°W; 20°S-25°S), with harmful consequences. Sao Paulo state is the most populous and economically important state not only for SEB, but for the entire country. The capital, Sao Paulo city, at 750 m above sea level, is the most populous city in the region, with an area of 1,521 km^2. According to the climate classification Köppen-Trewartha (K-T), the type of climate is given as Cwa[1] (humid subtropical climate). Figure 1 shows the location of Sao Paulo state. According to the National Institute of Meteorology[2] (INMET), September of 2015 was the rainiest September since 1993. The record precipitation value in 1993 was 206.7 mm, while in September 2015 it was registered as 199.9 mm. The mean annual precipitation for Sao Paulo city is 78.8 mm[3]. In the present paper, the atmospheric conditions responsible for the rainstorms that occurred in Sao Paulo state during 8th September 2015 and their consequences for the population, civil defence and electric energy distribution are analysed.

DATA AND METHODOLOGY

In order to analyse the adverse effects of the hostile meteorological conditions in many regions of Sao Paulo state, information from news agencies is used. Satellite images are used to identify atmospheric conditions responsible for the intense rainfall associated with strong winds on 8th September 2015. Data of surface pressure, vertical velocity and divergence in high levels are used to analyse the synoptic conditions. For this purpose, 6 hourly gridded datasets from the Climate Forecast System Reanalysis[5] (GFS4) with a horizontal resolution of 0.5° are used. The CAPE (Convective Available Potential Energy) Betts [10] was calculated using data from available from the University of Wyoming[6]. CAPE, effectively the positive buoyancy of an air parcel, is an indicator of atmospheric instability, which makes it very valuable in predicting severe weather. High values of CAPE are indicative of high atmospheric instability. Radar images[7] are also used to verify the atmospheric instability in the region.

[1]https://pt.wikipedia.org/wiki/Classificação_climática_de_Köppen-Geiger

[2]www.inmet.gov.br

[3]http://sao-paulo.estadao.com.br/noticias/geral,capital-paulista-registra-osetembro-mais-chuvoso-desde1993,1770753

[4]http://nomads.ncdc.noaa.gov/data/gfsanl

[5]http://nomads.ncdc.noaa.gov/data.php

[6]http://weather.uwyo.edu/upperair/sounding.html

[7]http://www.redemet.aer.mil.br/?i=produtos&p=radares-meteorologicos

In addition, data from meteorological stations of INMET in several cities in different regions of the Sao Paulo state are used to analyse the hourly variation of wind gust, minimum pressure and precipitation. The rainfall data in each one of these cities come from rain gauges. The meteorological stations record wind gust, minimum pressure and precipitation.

RESULTS

Synoptic conditions

The intensification of an area of low pressure located around Brazil, Bolivia and Paraguay together with a cold front passage from the southern region of Brazil to Sao Paulo and Rio de Janeiro States caused the development of several cumulonimbus clouds over the centre-south of Brazil, which subsequently provoked rainstorms (Figure 1). As seen in Figure 2, at 06:00 UTC, the low pressure region was located over Bolivia, Paraguay and the Mato Grosso do Sul state (westward from Sao Paulo) and moved from west to south at 12:00 UTC, 18:00 UTC and at 00:00 UTC reached Sao Paulo state. The strong and large squall line associated with the cold front over Sao Paulo state caused a high number of violent rainstorms in almost the entire state. Satellite images for 8th September show the strong activity during the afternoon in Sao Paulo state (Figure 1). Radar image at 20:00 UTC for Sao Roque[8] (near Sao Paulo city) shows the presence of rainstorms (around 90 mm) over the centre-east of Sao Paulo state (Figure 3), in agreement with Figure 1c.

[8]www.redemet.aer.mil.br/?i=produtos&p=radares-meteorologicos

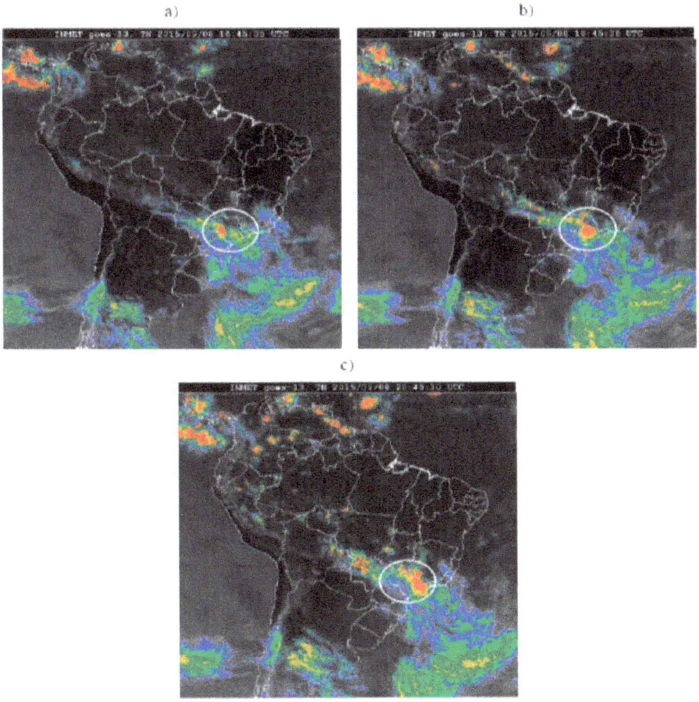

Figure 1: Infrared satellite image for 8th September 2015: (a) 16:45 UTC, (b) 18:45 UTC, 428 and (c) 20:45 UTC. The circle indicates the region of Sao Paulo state.

Strong divergence at upper levels associated with difluence (Figure 4) and intense ascending motion (Figure 5) due to the mesoscale convective cloud clusters were observed in the region. The displacement of the regions of strong divergence and intense vertical velocity (Figures 4 and 5) was in agreement with that of the sea level pressure as illustrated in Figure 2. As can be seen in Figure 4c-4d and Figure 5c-5d, areas of strong divergence and vertical velocity were present over Sao Paulo state at 18:00 UTC on 8th September and 00:00 UTC on 9th September. Figure 4b-4d also shows the presence of strong winds at 200 hPa in Sao Paulo state (around 25 m s^{-1}, i.e., higher than 90 km h^{-1}). This synoptic pattern is highly favourable for the occurrence of rainstorms and intense vertical movement that can intensify the winds close to the surface with harmful consequences, as happened in Sao Paulo state.

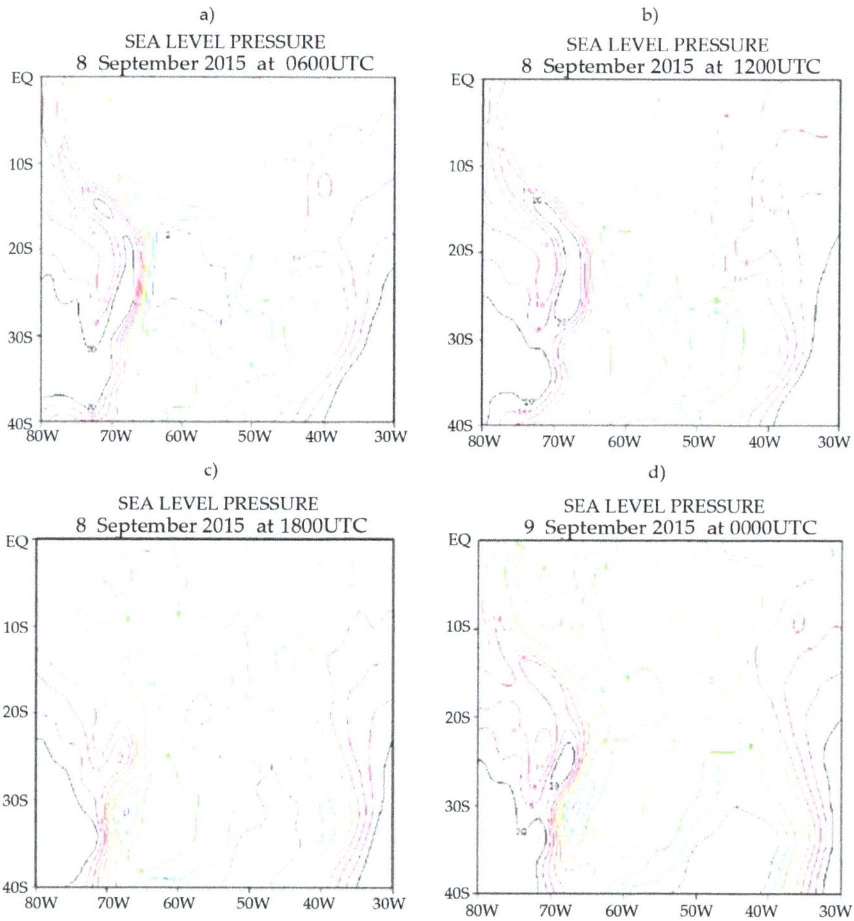

Figure 2: Sea level pressure for: (a) 06:00 UTC (b) 12:00 UTC (c) 18:00 UTC 04:32 on 8th September and (d) 00:00 UTC on 9th September 2015. Units: hPa.

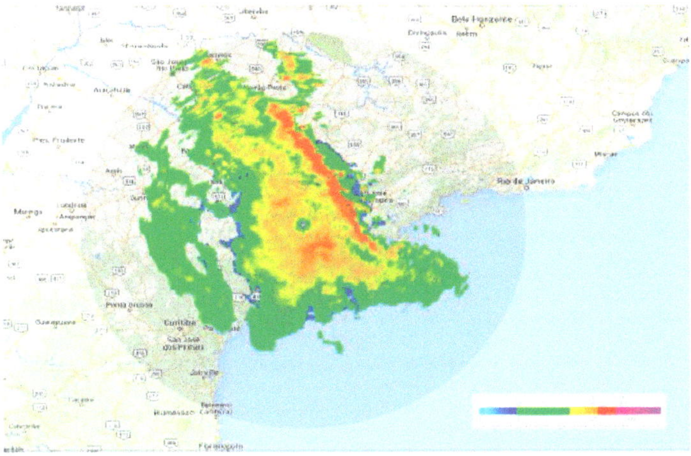

Figure 3: Radar image for 8th September 2015 at 20:00 UTC showing the rainstorm in the centre-east region of Sao Paulo state. The number in the upper side of the legend correspond to dbz and those in the lower side correspond to precipitation.

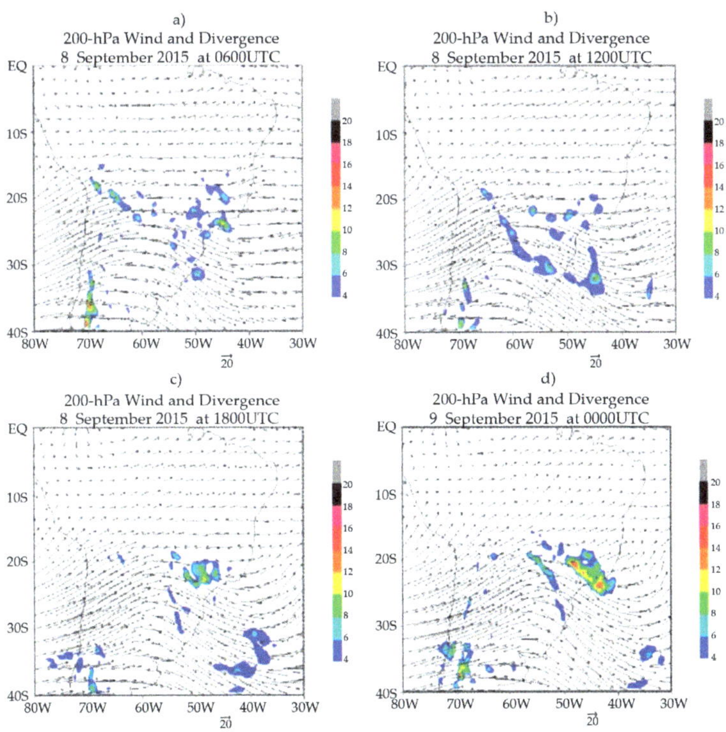

Figure 4: As in Figure 2, but for the divergence at 200 hPa (s^{-1}). Also shown are the wind vectors ($m\ s^{-1}$) at 200 hPa. Units: $m\ s^{-1}$.

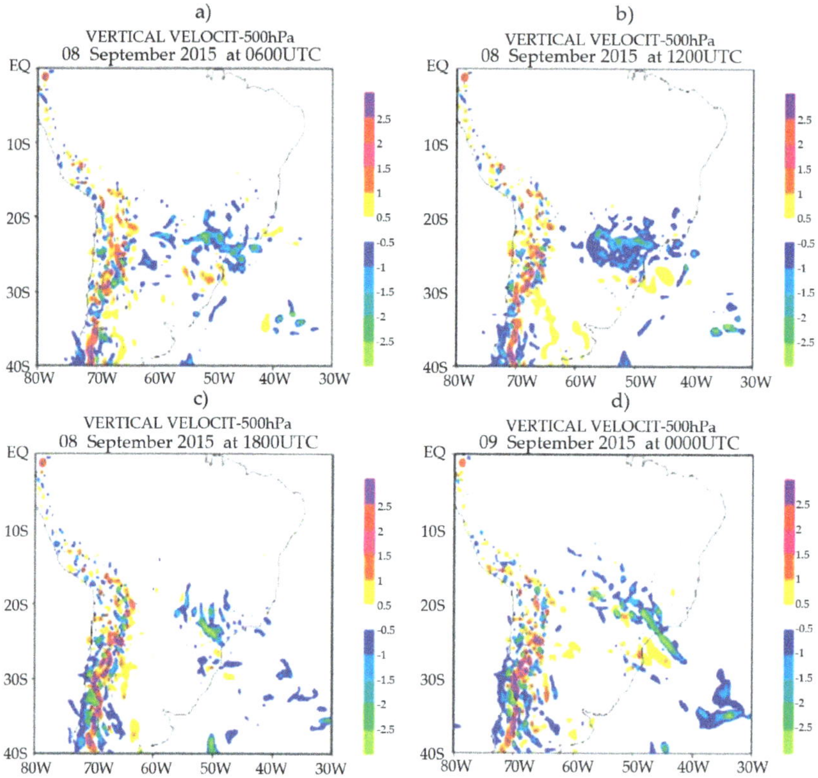

Figure 5: As in Figure 2, but for the vertical velocity at 500 hPa. Units: hPa s^{-1}.

Due to the large number and highly violent rainstorms, a vast amount of atmospheric discharges were observed in Sao Paulo state. According to the Atmospheric Electricity Group (ELAT) of the National Institute of Space Research (INPE), Brazil, 99,563 intra-cloud electric discharges and 31,929 cloud-to-ground discharges were observed during the 24 h of the 8th September 2015 (Figure 6). The large number of lightning events, together with strong winds and the falling of trees were responsible for un-roofing residences, the interruption of traffic on the streets and the electric system distribution.

In order to know the hourly values of wind gust, minimum surface pressure and precipitation during the 8th September 2015 in several cities in Sao Paulo state, figures 7-10 were prepared. In Figure 7, some cities in different areas of Sao Paulo state that have INMET meteorological stations were chosen. Although in Campinas and Limeira there are no INMET meteorological stations, their locations were inserted into this figure for use in the discussion of the results. In Figures 8-10, the hourly data of the wind gust, minima surface pressure and precipitation from 6th to 10th September 2015 in the cities of Figure 7 were showed. As can be noted, the stronger winds, the lower values of minima surface pressure and the heavier rainfall occurred during the 8th September 2015 in the afternoon, due to the presence of the strong squall line over

Sao Paulo state. In general, wind gust was higher than 18 m s^{-1} (approximately 65 km h^{-1}) and in some regions the wind reached 25 m s^{-1} (91 km h^{-1}). The hourly precipitation matches approximately the minima surface pressure. The rainfall was in general higher than 10 mm in almost the entire state; in the northwest of the state it reached 30 mm.

Figure 6: Electric discharges during 24 hrs for 8th September 2015 observed in Sao Paulo state.

Social and Economic Impacts

As show in Figure 1, on 8th September 2015 a strong and large squall line was present during the afternoon over Sao Paulo state. During 24 h, rainfall amounts of up to 100 mm occurred in almost all the areas of the state. This severe weather condition affected practically the entire Sao Paulo state and many cities suffered the impacts of the adverse meteorological conditions.

The strong squall line moved from the west to the east over Sao Paulo state, associated with a cold front (Figure 1). Using Python[9], a thermodynamic diagram was obtained where the CAPE value was calculated using radiosonde launched at 12:00 UTC in Campo Grande (Mato Grosso do Sul State, westward of Sao Paulo). Campo Grande is the nearest location to Presidente Prudente (located in the west of the state, see Figure 7) with atmospheric sounding registers. The CAPE presented a value higher than 1500 J kg^{-1} showing atmospheric instability in the region (Figure 11). Unfortunately, there were no more atmospheric soundings during the day in Campo Grande. However, taking into account that the CAPE presented a high value already at 12:00 UTC, it can be inferred that the atmospheric instability increased during the development of the observed mesoscale convective clusters in the squall line in the west of the Sao Paulo state.

At 16:45 UTC, intense convective activity was observed over the west region of Sao Paulo state (Figure 1a). The impact on the cities located in this region was strongly destructive. Figure 12a illustrates the effect of the heavy rainfall and strong winds in the region of Presidente Prudente. Due to the rainstorm, many places were flooded causing the floating of cars. Many trees fell down and the traffic of vehicles was affected[10]. At 18:45 UTC, the strong convective activity moved to the central region of the state (Figure 1b). In Barra Bonita (Figure 7) many areas were flooded due to the intense rainfall and vehicles were dragged by the water current[11] (Figure 12b).

[9]https://pypi.python.org/pypi/SkewT

[10]http://www.imparcial.com.br/site/forte-chuva-alaga-diversos-pontos-de-prudente-source:e-provoca-estragos.html

[11]http://noticias.bol.uol.com.br/ultimas-noticias/brasil/2015/09/30/enxurradacarrega-carros-e-motos-em-rua-principal-de-barra-bonita-sp.htm

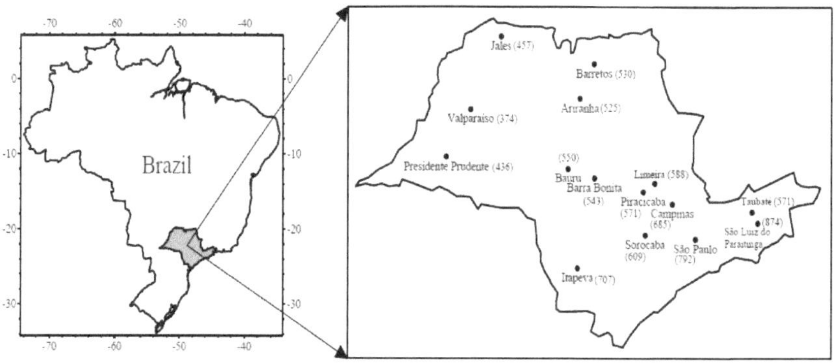

Figure 7: Location of some cities in different regions of Sao Paulo state. All have meteorological stations of INMET, apart from Campinas and Limeira.

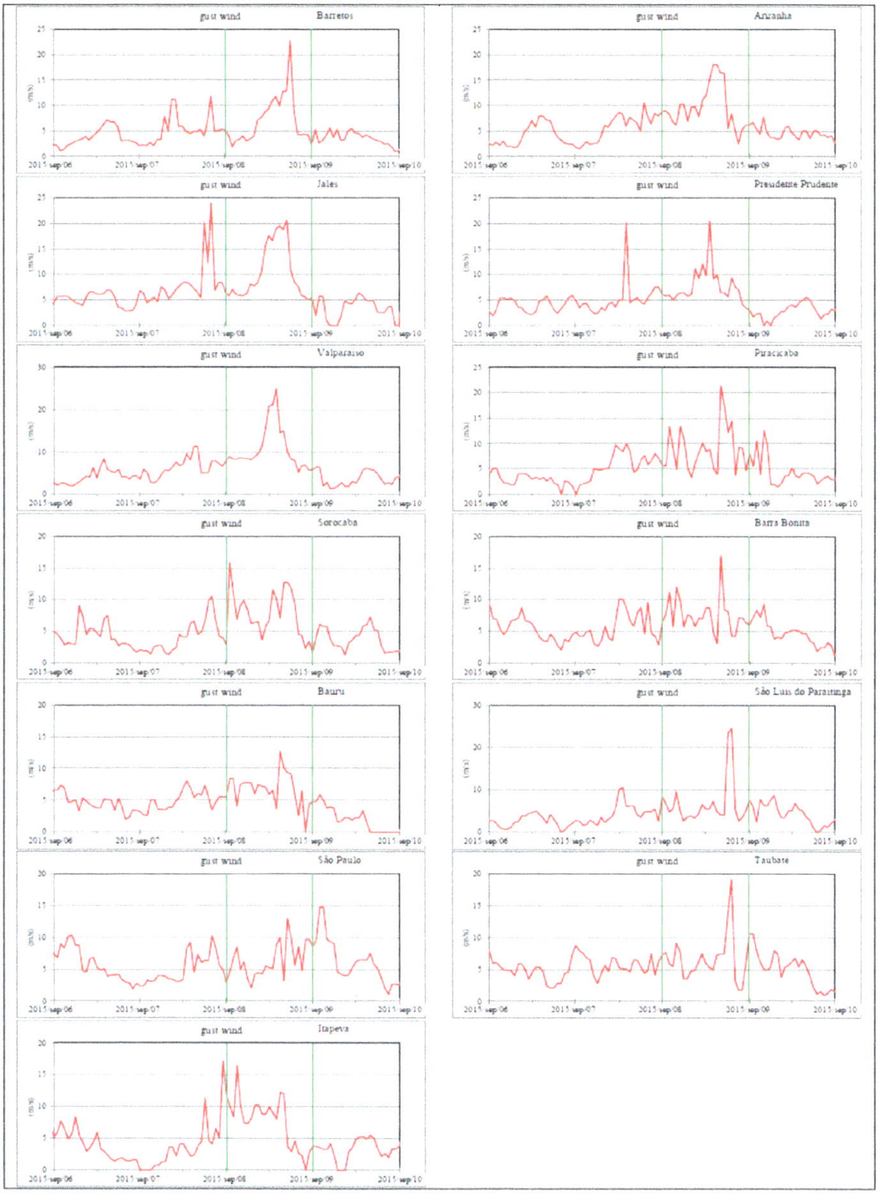

Figure 8: Hourly data of wind gust in the cities of Figure 7 that have meteorological stations of INMET (Units: ms^{-1}). The green lines delimit the day 8th September 2015.

Figure 9: The same as in Figure 8, but for the minima surface pressure (Units: hPa).

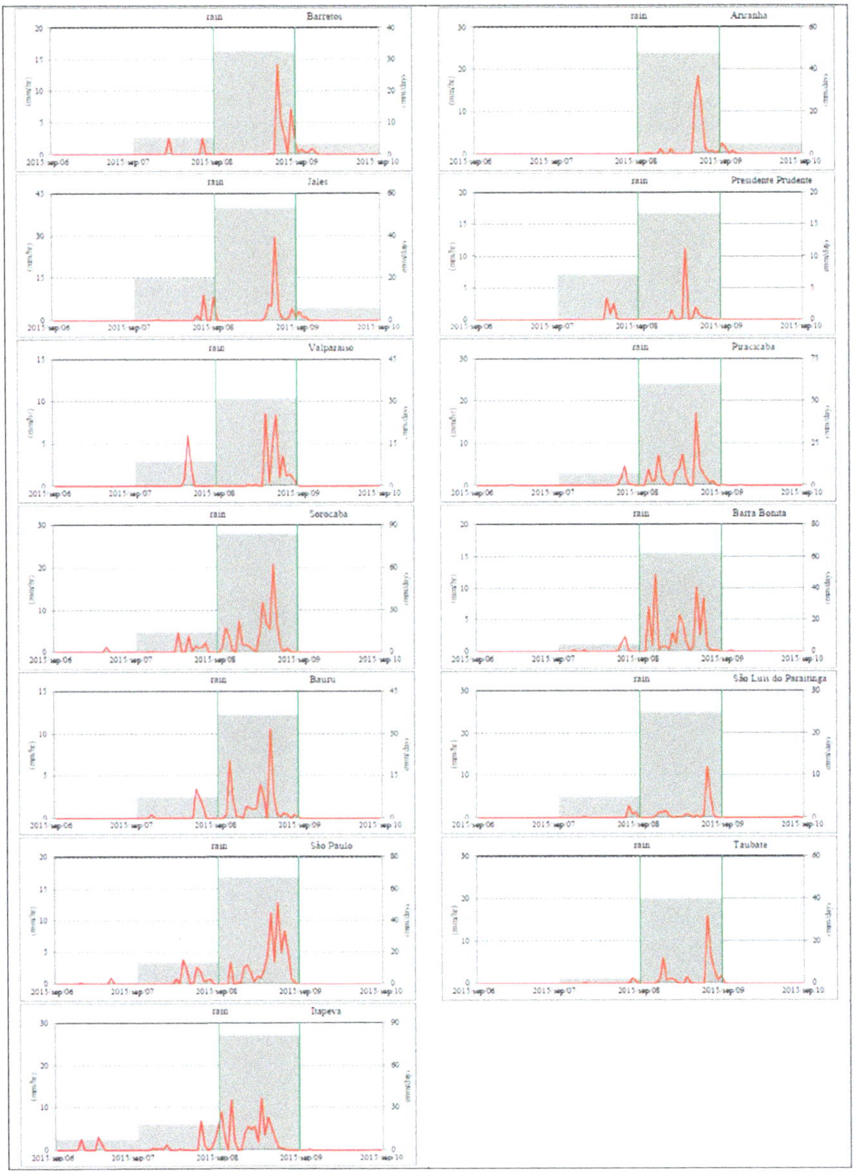

Figure 10: The same as in Figure 7, but for precipitation (Units: mm). Also, the accumulated precipitation is indicated.

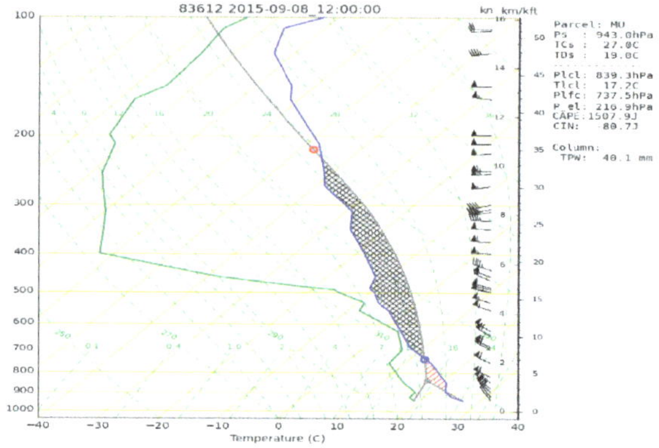

Figure 11: CAPE obtained at 12:00 UTC for Campo Grande (Mato Grosso do Sul state), westward from Sao Paulo state. The hachured area corresponds to the CAPE.

Figure 12: Damages due to the storm of 8th September 2015 in Sao Paulo state: (a) Presidente Prudente, (b) Barra Bonita, (c) Campinas, (d) Limeira, (e) Piracicaba and (f) Sao Paulo.

Several problems occurred in the region of Campinas (Figure 7), an important city in the centre-east region of Sao Paulo state. At least 736,000 people remained without electric energy due to strong wind and lightning associated to the rainstorm which damaged the electric distribution system[12]. In addition, flooding and the falling of trees was observed. The wind reached 98.3 km h^{-1} and the civil defence stated that 26 trees fell down. Figure 12c shows the onset of the rainstorm in Campinas[13]. Many other cities in the centre-east sector of Sao Paulo state were affected by rainstorms. Figures 12d-12e illustrate, respectively, the onset of the storm in Limeira[14] and the damage in Piracicaba, where cars were dragged by the water, trees were fallen down and an electric cut was observed in many points of the city[15] (Figure 7).

At 20:45 UTC, a strong convection was present in the east sector of Sao Paulo state (Figure 1c). In Sao Paulo city, the rainfall during 8th September 2015 was the highest for September since the last 20 years. The amount of precipitation (69.3 mm) was 3% higher than the amount expected for the entire September (67 mm). At around 21:30 UTC, there were 29 regions with inundation, 5 of which were impassable. The traffic was interrupted in many regions of the city, the metropolitan trains and some subway lines showed problems due to the strong rainfall. At 21:30 UTC Sao Paulo city[16] showed 184 km of traffic jams. The strongest rainfall occurred in the west region of Sao Paulo city (Figure 13). The accumulated precipitation from 23:00 UTC on 7th September 2015 to 23:00 UTC on 8th September 2015 in Butantan district was 97.8 mm. Wind gusts of 83 km h^{-1} occurred in the region. Due to the storm, the distribution of the electricity system suffered, the worst damage seen that year, and at least 149 trees fell down in the city. Figure 12f illustrates the harm in Sao Paulo city[17].

The metropolitan region of Sao Paulo has a population of 20 million. Each region of inundation in Sao Paulo city causes a daily damage with costs higher than US$ 263,000,00 (1,000,000 Reais). As there were almost 749 regions of inundation, the annual losses reached approximately US$ 88.4 million (762,000,000 Reais) [11]. The inundation reduces the city development and quality of life for the inhabitants. Direct losses are due to the interruption of economic activities in the areas affected by floods. In addition, the floods affect daily traffic inside and outside the metropolitan region.

In the present study, the rainstorms only on 8th September 2015 in Sao Paulo state were considered. However, September 2015 was an atypical month where many rainstorms occurred, not only in Sao Paulo state but also in the centre-south of Brazil which deserves future research.

[12]http://www.cpfl.com.br/releases/Paginas/temporal89.aspx

[13]http://g1.globo.com/sp/campinas-regiao/noticia/2015/09/temporal-deixa-274-mil-sem-

energiaem-3-cidades-da-regiao-de-campinas.html

[14]http://g1.globo.com/sp/piracicaba-regiao/noticia/2015/09/imagem-mostra-mar-de-nuvens-no-ceu-de-limeira-e-vira-sucesso-na-web.html

[15]http://g1.globo.com/sp/piracicaba-regiao/noticia/2015/09/chuva-de-20-minutos-alaga-avenidas-e-arrasta-carro-no-centro-de-piracicaba.html

[16]http://sao-paulo.estadao.com.br/noticias/geral,chuva-em-sp-faz-corrego-transbordar-evento-derruba-arvores,1758465

[17]http://sao-paulo.estadao.com.br/noticias/geral,capital-paulista-registra-o-setembro-maischuvoso-desde-1993,1770753

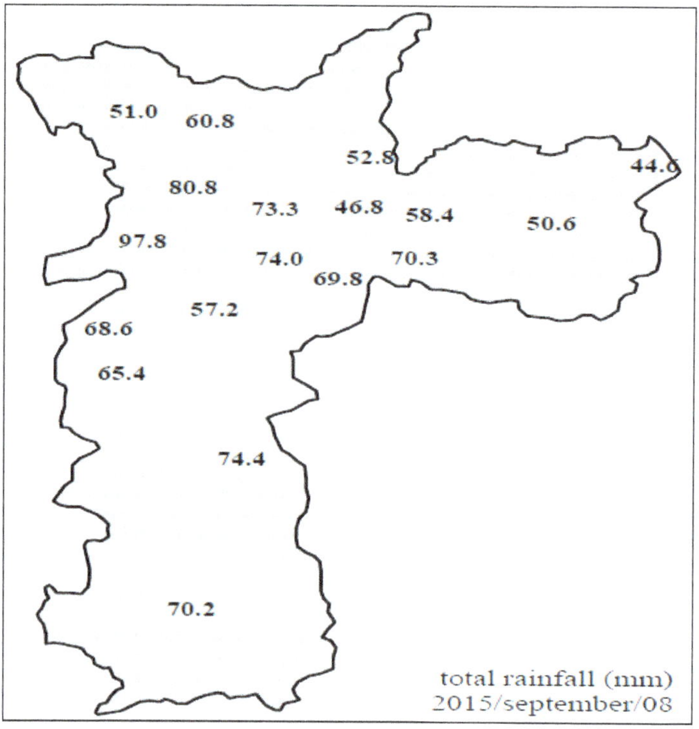

Figure 13: Accumulated rainfall in the regions of Sao Paulo city.

Mitigation efforts to severe weather conditions

Extreme weather events like that of 8th September 2015 seem to be occurring more and more frequently, often with disastrous results. This is aggravated by the evidences reported in the AR4 and AR5 IPCC (Assessment Report 4 and 5 of the Inter-governamental Panel on Climate Change), which indicated that in the last decade extreme events such as tornadoes, heat waves and strong thunderstorms associated with rainstorms, windstorm and high incidence of lightning are occurring more frequently. In particular, in Brazil, the extreme events responsible for natural hazards have been occurring in almost all the regions, for example those that occurred on September 2015, the Catarina cyclone [12,13], which left 33,000 people homeless in the State of Santa Catarina. Also the strong rainstorms which occurred in Angra dos Reis and many other regions of the State of Rio de Janeiro in 2010 caused more than 200 regions of landscape and the death of more than 170 people, the rainstorms in the mountainous regions of

Sao Paulo and Rio de Janeiro States in 2011 [14] which provoked the death of more than 900 people and destroyed thousands of homes and the recent thunderstorms of November 2015 in the centre-south of Brazil[18]. Table 1 shows some of the extreme events occurring in Brazil and their consequences. Thus, there is need to mitigate the effects of the severe extreme events. Studies providing information of the atmospheric conditions in extreme situations are essential for the electric energy system security, national economy and the population comfort.

[18]http://www.deolhonotempo.com.br/index.php/nacional/3412-tres-tornados-saoregistrados-num-so-dia-no-brasil

In addition, it is necessary to have a system of analysis and meteorological management that allows: a) ample access to meteorological data aggregating the network of meteorological, hydrological, radar and satellite data, b) to locally forecast the severe rainstorms at least 6 h ahead of time, c) to enable a warning scheme.

Recently, the project named "Management of the impact of climate severe events on the electric energy network" (In Portuguese, "Gestão dos impactos de eventos climáticos severos na rede elétrica") has been developed with the purposes mentioned above. It is a co-operative project between the National Institute for Space Research (INPE) and Energisa. It is an interdisciplinary project with the participation of the ELAT/INPE Group, which develops studies of thunderstorms and electrical discharges, the Centre of Weather Prediction and Climate Studies (CPTEC/INPE) which run numerical weather prediction models and makes available several products of environmental monitoring for the society, the Division of Image Processing (DPI/INPE) which has developed the TerraMA2 Library that enables the generation of geo-processing of spatial data in management systems.

The warning system is based on a platform that allow us to integrate in real time geo-environmental data (atmospheric, climatological, hydrological, geotechnical) with maps of environmental risks and vulnerabilities in order to give warning in cases such as floods, dry weather and electric discharges, among others. This platform is called TerraMA2 – Monitoring, Analysis and Alert Plataform[19]. The TerraMA2 needs essentially of data obtained from the geo-environmental module and the mapping of the areas of risk. The geo-environmental data provides information on weather, climate and other extreme events obtained from analysis of observational data and numerical weather predictions. The module that stores information about the risks and vulnerability associated to natural disasters takes into account geotechnical information obtained from local research institutes. The information associated with climate and hydro-meteorological extremes are combined with the mapping of the areas of potential risks. The intersection of the information allows us to identify potential risks which supplies the module of automatic analysis. The automatic analysis is converted to warnings that are sent to the agencies responsible for prevention and mitigation. As well as being important for the electricity sector, the results of the project can be applied to other sectors such as agriculture and civil defence, among others. The project has just begun and future case studies will be presented in another papers. Studies such as those in the present paper provide information on climate extremes, which is useful to the project.

SUMMARY

In this paper, the atmospheric conditions of the strong rainstorms which occurred in the Sao Paulo state during the 8th September 2015, as well as the social and economic damage, were analysed. This severe weather condition affected practically all the areas of the state and many cities suffered the impacts of the rainstorms. Intense rainfall associated with strong winds and lightning caused floods and threw down trees, turned street lamps out, unroofed residences, interrupted traffic on the streets and dropped conducting wires affecting the transmission and distribution of electric energy, all causing damage for the population. In the present study the harmful effects of the rainstorms in various cities in different regions of Sao Paulo State were illustrated and discussed. Hourly data of wind gust, minima surface pressure and precipitation from the 6th to 10th September 2015 in several cities of the state were plotted. It was observed that during the 8th September, the wind gust was in general higher than 18 m s^{-1} (approximately 65 km h^{-1}); in some regions the wind reached 25 m s^{-1} (91 km h^{-1}). The hourly precipitation matches approximately the minima surface pressure. The rainfall was in general higher than 10 mm in almost the entire state; in the northwest of the state it reached 30 mm.

[19]www.dpi.inpe.br/terrama2

Table 1: Extreme events over Brazil and their consequences

Extreme event	Consequences
Catarina cyclone in 2004	left 33,000 people homeless in the State of Santa Catarina
strong rainstorms in Angra dos Reis in 2010	caused more than 200 regions of landscape and the death of more than 170 people
rainstorms in the mountainous regions of Sao Paulo and Rio de Janeiro States in 2011	provoked the death of more than 900 people and destroyed thousands of homes
Rainstorms in Sao Paulo State in September 2015	caused floods and threw down trees, turned street lamps out, unroofed residences, interrupted traffic on the streets, dropped conducting wires affecting the transmission and distribution of electric energy producing damage for the people
thunderstorms that happened in November 2015 in the center-south Brazil	In Chapeco (State of Santa Catarina): entire buildings destroyed in few seconds due to strong winds; In Marechal Candido Rondon (State of Parana): the tornado destroyed part of the city and caused the death of 10 people

The storm over Sao Paulo state was caused by a large and strong squall line moving from west to east in the region. Low pressure associated with strong divergence at high levels and ascending motion was observed. Satellite and radar images were used to illustrate the strong convective activity in the region, particularly at 20:00 UTC in the centre-east sector (heavier rainfall around 90 mm). Due to the large number and the high violent rainstorms, a fantastic amount of atmospheric discharges was observed. Data from ELAT/INPE indicated that 99,563 intra-cloud electric discharges and 31,929 cloud-to-ground discharges were observed during the 24 hrs of the 8th September 2015.

Particularly in Sao Paulo city, the rainfall during the 8th September 2015 was the highest registered during September in the last 20 years. The amount of precipitation (69.3 mm) was 3% higher than the amount expected for the entire September (67 mm). Each region of inundation in Sao Paulo city causes a daily damage higher than US$ 263,000,00. Since the storm of the 8th September affected almost all the regions of the city, a great economic burden is expected. A co-operative project between the INPE and Energisa is in progress, aimed to mitigate the effects of adverse weather conditions as in the case of the rainstorm during the 8th September 2015. The project has just begun and it is expected that the results can help the electric energy system, the national economy and the population comfort.

ACKNOWLEDGEMENTS

This study forms part of the project "Management of the impact of climate severe events on the electric energy network" financed by the Energisa power company.

REFERENCES

1. 1. Zhou J, Lau KM (1998) Does a monsoon climate exist over South America? J Climate 11: 1020-1040.
2. 2. Gan MA, Kousky VE, Ropelwski CF (2004) The South America monsoon circulation and its relationship to rainfall over west-central Brazil. J Climate 17: 47-66.
3. 3. Vera C, Higgins W, Amador J, Ambrizzi T, Garreaud R (2006) A unified view of the american monsoon systems. J Climate 19: 4977-5000.
4. 4. Rao VB, Franchito SH, Gan MA, Gerolamo ROP (2013) Duration of the South American summer monsoon is increasing. Atmos Sci Letters 15: 110-113.
5. 5. Massambani OL, Vespoli MV, Vazquez MA (1992) Tornado ou microexplosão? Um diagnóstico via Radar do evento de Itú - São Paulo. Sociedade Brasileira de Meteorologia 2: 763-768.
6. 6. Lima KC, Satyamurty P, Fernandez JPR (2010) Large-scale atmospheric conditions associated with heavy rainfall episodes in Southeast Brazil. Theor Appl Climatol 101: 121-135.
7. 7. Seluchi ME, Chou SC, Gramani M (2011) A case study of a winter heavy rainfall event over the Serra do Mar in Brazil. Geofis International 50: 67-80.
8. 8. Satyamurty P, Silva MT (2011) Trends in the frequency of intense precipitation events in Southern and Southeastern Brazil during 1960-2004. J Climate 24: 1913-1921.
9. 9. Sias EEK, Dias MAF (2002) Análise de um caso de precipitação extrema no inverno na cidade de São Paulo. Congresso Brasileiro de Meteorologia.
10. 10. Betts AK (1974) Thermodynamic classification of tropical convective sounding. Mon Wea Rev 108: 760-764.
11. 11. Haddad EA, Teixeira E (2015) Economic impacts of natural disasters in megacities: A case of floods in Sao Paulo. Habitat International 45: 106-113.
12. 12. Cowan RM, Bosart LF, Davis CA, Atallah EH, Gyakum JR et al. (2006) Analysis of Hurricane Catarina (2004). Mon Wea Rev 134: 3029-3053.
13. 13. *Pezza AB, Simmonds I (2006)* Catarina: The first South Atlantic hurricane and its association with vertical wind shear and high latitude blocking. *8th International Conference on Southern Hemisphere Meteorology and Oceanography. Instituto Nacional de Pesquisas Espaciais pp: 353-364.*

14. 14. Netto ALC, Sato AM, Avelar AS, Vianna LGG, Araújo IS et al. (2011) January 2011: the extreme landslide disaster in Brazil. Proceedings of the Second World Landslide Forum, Italy.

13

CHAPTER

IMPACTS OF URBANIZATION ON WATER RESOURCES AND VEGETATION ON THE DELTA REGION OF TAMILNADU USING REMOTE SENSING AND GIS

Arunpandiyan M, Aarthi M, Vidyalakshmi R, Savaridhos RJ and Prashanthi Devi

Keywords: Gender, Women, rural area, agriculture, policy, program.

INTRODUCTION

Water occupies a unique place of all of the planet's renewable energy sources. The water resources of a country are one of its vital assets [1]. Properly managed water resources are a critical component for all forms of life, food production, economic growth, poverty reduction and general wellbeing. The livelihoods of the poorest are critically associated with access to water services. It is impossible to substitute, difficult to de-pollute, expensive to transport, and it is truly a unique gift to mankind from nature. Water is also one of the most manageable of the natural resources as it is capable of diversion, transport, storage, and recycling. The surface water and groundwater resources of the country play a major role in agriculture, hydropower generation, livestock production, industrial activities, forestry, fisheries, navigation, recreational activities, etc. According to the National Water Policy in the planning and operation of systems, water allocation priorities are recommended as: (i) drinking water, (ii) irrigation, (iii) hydropower, (iv) ecology, (v) agro-industries and non-agricultural industries, and (vi) navigation.

Anthropogenic activities are of a major force in impacting the hydrological cycle, as well as the climate. Assessing the impacts of human activities on hydrological environments is becoming a wide-focused topic. Accelerated groundwater exploitation over the past few decades has resulted in great social and economic benefits by providing low-cost, drought-reliable and high quality water supplies for urban areas, rural populations and crop irrigation. The rapidly increasing large cities in semiarid and semi-humid regions raise many problems within water resources management

Citation: Arunpandiyan M, Aarthi M, Vidyalakshmi R, Savaridhos RJ and Prashanthi Devi, Impacts of Urbanization on Water Resources and Vegetation on the Delta Region of Tamilnadu Using Remote Sensing and GIS, ISSN:2347-7822

The Cauvery Delta region, i.e. the Thanjavur, Tiruchirappalli and Tiruvarur districts, recorded deficit rainfalls from -3 to -61%, with an averaged deficit of approximately -16.8% over the past four years. Though this deficit has recently occurred due to fast urbanization, other reasons such as climate change, lack of monsoons and global warming are also significant. Out of these, urbanization plays an important role in the depletion of water resources. In the last 50 years, urban growth has led to large scale destruction of this physical heritage of study area. Large scale encroachments have led to the filling up of lakebeds and the conversion of built-up areas, by both the government and private agencies over the last few decades. Land clearing, primarily for agriculture in earlier days, but recently due to urbanization, is perhaps the single most important cause of environmental degradation, loss of species, and depletion of ecological communities. Forest loss and degradation are particular problems for conservation biology because forests are some of the most species-rich environments on the planet [2], particularly for birds [3] and invertebrates [4]. The United Nations Food and Agriculture Organization (FAO) reviewed global forest clearance, and found that between 1990 and 2000, the average rate of forest loss was 16.1 million hectares per year [5]. The causes of desertification and vegetation destruction are usually grouped into natural and human causes. Desertification has been attributed to long-term climatic changes, cyclic fluctuations in climate, periodic droughts and destructive consequences of human activities

In an arid environment, the main limiting factor for plant production is the amount of water in the soil available for the vegetation. Droughts are common in arid regions and occurs when a region has insufficient moisture to meet the demands of the plants, people, livestock or wildlife. A drought is meteorological when rains fail to reach a certain level over a particular period of time in a given area, whereas agricultural drought occurs when the combined effects of amounts and distribution of rainfall, soil water reserves, and evaporation bring about a drastic reduction of agricultural yields and livestock, leading to food scarcity and other associated problems. Hydrological drought is brought about due to changes that reduce the absorption and storage of moisture in the soil. The objective in this present study is to analyse the extent to which urbanization has affected the small water resources in this area, and to provide appropriate and affordable strategies for conserving the water resources and quality in this community.

STUDY AREA DESCRIPTION

The study area is made up of the Tiruchirapalli, Thanjavur and Tiruvarur regions of Tamil Nadu (Table 1). The total geographic study area covers approximately 10177 km^2. The district of Tiruchirapalli consists of 9 taluks (administrative divisions), the Thanjavur district 8 taluks and Tiruvarur 7 taluks. The districts are surrounded by Pudukottai, Karur, Namakkal, Perambalur, Ariyalur, Nagapattinam district as per clock wise directions and the Bay of Bengal on the south east (Figure 1).

Table 1: Water resources potential of basin-surface water

S. No.	Major River Basins	Minor River Basins	Water Resources potential of basin-Surface Water
1	Cauvery	Vennar Vettar Kudamurutti Koraiyaru Odampoki	7067

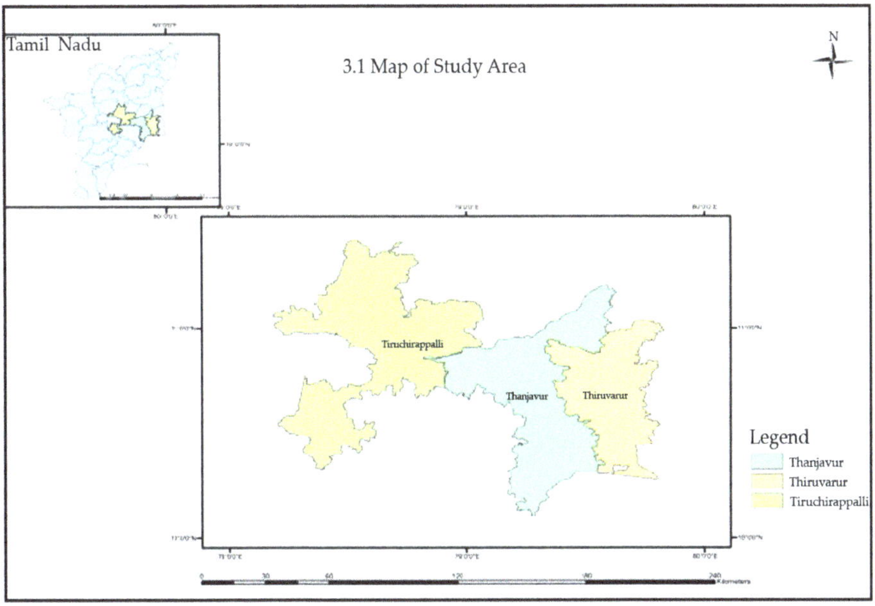

Figure 1: Study area location.

DATA SOURCES AND METHODOLOGY

The data used for the present study includes the Survey of India topographic sheets of the year 1972, at the scale of 1:50,000. These consist of 34 maps covering the entire study area. The IRS LISS III images for the study area were downloaded through the portal ISRO Bhuvan and used for further processing. The near-infrared, red and green bands were layer stacked to obtain a false colour composite. The images were classified using a supervised classification (maximum likelihood) into four classes: urban settlements, drainages, vegetation and water body. Change detection analysis was performed to detect the rates of depletion among water and vegetative resources. The methodology flowchart is presented as Figure 2.

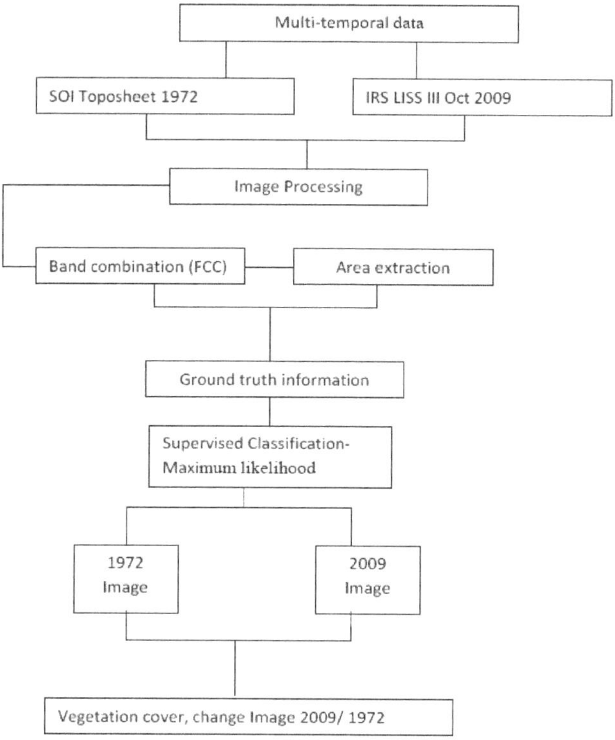

Figure 2: Methodology flowchart.

RESULTS

The surface of the land is an important resource that is constantly modified by human activity. Rapid population expansion has placed great demands for living space, leading to increasing urbanization and the exploitation of marginal lands. The latter has led to problems in many tropical and subtropical areas, such as deforestation, land degradation and desertification, flooding. In cold regions, the landscape is particularly sensitive and disturbances to the thermal equilibrium of permafrost has led to the development of thermokarst. The increasing pace of urbanization has also led to major engineering works on slopes and coastal reclamation. The former can lead to increased landslide hazards, while the latter usually occurs at the expense of coastal fisheries and ecologically important wetlands. Urbanization leads to changes in the hydrology, geomorphology, water quality and riparian vegetation of rivers, leading to degraded sites and reduced biodiversity. The comparison of the multi-temporal GIS data shows the depletion of water bodies in the study area over a period of 37 years. From the results, the estimation of the water bodies is 176 km^2. Similarly, the urban area or area under construction was observed to be 227 km^2 (Figure 3).

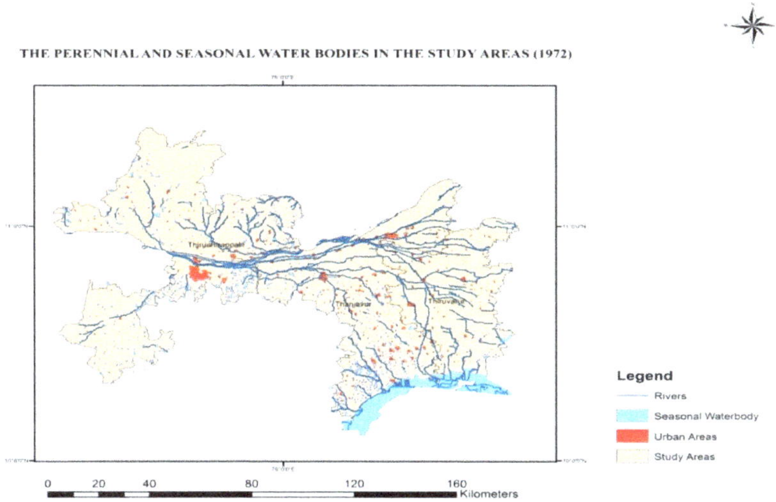

Figure 3: Perennial and seasonal water bodies in the study area.

The IRS LISS III image for the year 2009 was subject to supervised classification using a visual interpretation method. Land classes, such as vegetation, urban areas and water bodies were extracted from the supervised classes and the area in km^2 was estimated for each class (Figure 4). From the results, it was observed that the water body area covered 68 km^2, while urban areas covered 649 km^2. The observed results are presented in Table 2.

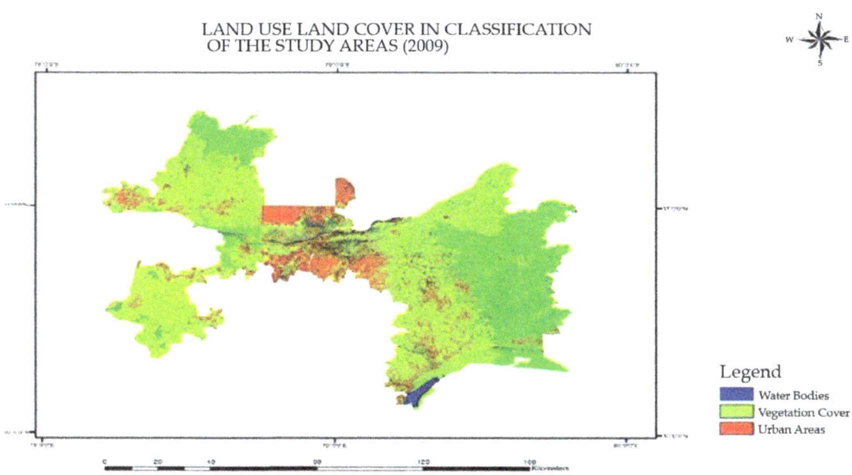

Figure 4: Land use land cover classification of the study area (2009).

LULC Class	Year 1972 Area in km²	Year 2009 Area in km²	Changes in Area (km²)
Urban	226.9	648.51	421.61 (+)
Water bodies	175.943	68.0982	107.844 (-)
Vegetation	9492.25	3418.47	6075.78 (-)

Table 2: Changes in land use/ land cover in study area.

A comparison of the areas estimated for urban and water bodies was carried out to identify land use changes over the study period of 37 years. Results in Table 2 show that the urban areas increased by 422 km^2 and water bodies drastically decreased by 108 km^2. A large decrease was also observed for land under vegetation, at 6076 km^2. To evaluate the depletion of vegetation cover, the same procedure was adopted. It was estimated that the area under vegetation cover was 9492 km^2 during 1972, while urban areas or areas under construction was observed at 227 km^2. For the year 2009, vegetation cover had fell to 3418 km^2 and the urban areas increased to 649 km^2.

CONCLUSION

From the present study, it is revealed that the natural resources within the study areas have experienced a strong decline over the period of 37 years (1972-2009) due to urbanization activities. This value exceeds the naturally estimated decline for the same region. Hence, conservation measures have to be adopted for the optimal utilization of these resources without compromising their economic benefits and sustainability.

REFERENCES

1. 1. Rakesh Kumar, et al. Water resources of India, Current Science 2005; 89: 794-811.
2. 2. Civco DL., et al. A Comparison of land use and land cover change detection methods. ASPRS–ACSM Annual Conference and FIG XX11 Congress, Washington, DC, USA 2002; 22–26.
3. 3. Alcamo J, et al. A grand challenge for freshwater research: Understanding the global water system, Environ. Res. Lett., 2008; 3.
4. 4. Meyer JR. The Role of Industrial and Post Industrial Cities in Economic Organization, 2000.
5. 5. Briscoe, J. India's Water Economy. Bracing for a Turbulent Future. ReportNo. 34750- IN, viii–xi, World Bank, 2005.

14
CHAPTER

LANDSAT IMAGE ANALYSIS OF TREE MORTALITY IN THE SOUTHERN SIERRA NEVADA REGION OF CALIFORNIA DURING THE 2013-2015 DROUGHT

Christopher S Potter

Keywords: Landsat, Forest, Normalized difference vegetation index (NDVI), NDMI, Drought, Disturbance, Sierra Nevada, California

INTRODUCTION

Drought impacts on the vigor and long-term survival of forests across the state of California have become a pressing concern among scientists, water supply agencies, and land managers. Alarming projections of future tree mortality across the western United States have been linked to potential transforms of ecosystem structure, tree species composition, loss of biodiversity, and increased wildfire risks [1].

In April 2015, the United States Forest Service (USFS) flew more than 4.1 million forested acres in the southern Sierra Nevada region of California, with aerial surveys covering the low elevation western portions of the Stanislaus, Sierra and Sequoia National Forests (NF) and Yosemite and Sequoia-Kings Canyon National Parks [2]. The locations of dead and dying trees were recorded visually by a surveyor, using a digital sketch-mapping system on a fixed-wing aircraft flying approximately 330 meters above ground level. The surveyor recorded the number and species of affected trees and type of damage (mortality, defoliation, etc.) at each mapped location. These surveys mapped the locations of 10.45 million recently dead trees in the region. In the Sierra foothills, tree mortality was widespread, especially in ponderosa pine, gray pine, and potentially in blue oak and live oak. On the Stanislaus NF, mortality was scattered in northern areas, but roughly doubled since July 2014 in extensive ponderosa and other pine forests in the southern low-elevations areas. At higher elevations, conifer mortality

Citation: Potter CS (2016), Landsat Image Analysis of Tree Mortality in the Southern Sierra Nevada Region of California during the 2013-2015 Drought. J Earth Sci Clim Change. 7: 342, doi:10.4172/2157-7617.1000342

was scattered. On the Sierra and Sequoia NFs, tree mortality, mostly from western pine beetle, was common almost everywhere at lower elevations [2].

Satellite remote sensing has been shown to be an effective method to measure large-scale patterns of vegetation biomass and tree growth rates in remote mountain areas [3-11]. Van Wagtendonk and Root [12] reported that the satellite normalized difference vegetation index (NDVI) can provide information necessary to distinguish vegetation types when analysed over a summer growing season in the Sierra Nevada. Similar studies have found that NDVI is sensitive to early (herbaceous) post-fire recovery and subsequent woody (tree and shrub) regrowth trajectories [13,14]. Schiffman [15] reported a strong correlation ($R^2 = 0.7$) between July Landsat NDVI and leaf area index (LAI) measured in 175 forest stands of Yosemite National Park. LAI is an important indicator of forest stand condition, chiefly because of its close relation to canopy photosynthesis and transpiration rates [16]. Comparisons with field measurements have shown that remotely sensed NDVI from surface reflectance was significantly correlated with both LAI and with annual growth increments in coniferous forests and woodlands stands, including those in the western United States [17-20], and did not saturate until LAI levels were far greater than those commonly reported for forest stands of the subalpine Sierra Nevada.

In a regional study of annual moisture availability and forest greenness variations over 25 years, Trujillo et al. found a strong positive relationship between Sierra Nevada forest NDVI and maximum annual snow accumulation. The correlation between annual peak (July - August) satellite NDVI (at 8-km spatial resolution) and peak snow water equivalents (SWE) for evergreen needle leaf forests in the northern and central Sierra Nevada was between an R^2 of 0.7 and 0.8 over the period of 1982 to 2006. This study implied that the level to which snow accumulation can explain inter-annual variations in Sierra forest green cover is highest in the elevation zone between 2000 and 2600 m.

For additional mapping of drought impacts on forest stands, the Landsat normalized difference moisture index [21] has been validated in the field repeatedly as an index of vegetation canopy moisture status [22-24], including impacts of insect outbreaks on western conifer forests [25]. Potter [26] and Van Gunst et al. [27] showed that the NDMI can be used to monitor and map year-to-year variations in drought stress across California forest and woodland ecosystems.

The objective of the present study was to analyse 30 consecutive years of Landsat NDVI and NDMI values (at 30-m spatial resolution) at USFS aerial survey locations in the southern Sierra forest that showed high levels of tree mortality in April 2015, to gain a better understanding of what yearly satellite observations can track in terms of changes in stand structure and forest canopy variations. The year 2013 was the driest on record in California, with a total of just 30% of average state-wide precipitation [28]. A record low snowpack level in 2015 in the Sierra Nevada Mountains was unprecedented in comparison to the past 500 years [29], with SWE levels measured across the Sierra region for only 5% of the historical average. Landsat's unique three-decade record of vegetation cover change provides a rich archive, within which to put the 2013-2015 drought into historical perspective, particularly with respect to the impacts of observed tree mortality on stand canopy density, drought stress, and LAI.

STUDY AREA

The Sierra Nevada region covers over 63,000 km^2 along a north-south axis of California. Rundel et al. [30] defined elevation zones in the Sierra-Nevada roughly as: montane (1500 - 2500 m), subalpine (2500 - 3000 m), and alpine (above 3000 m). Before human settlements in the mid-1800s, conifer forests throughout the region experienced wildfires mainly of low and moderate severity. Montane zone pine and mixed conifer forests historically experienced mean fire return intervals of 11 to 16 years [31]. In higher elevation subalpine forests of the Sierra Nevada, the typical pre-settlement fire return intervals were between 35-100 years [32].

Among the most common tree species in the montane zone are ponderosa pine (*Pinus ponderosa*), Douglas fir (*Pseudotsuga menziesii*), white fir (*Abies concolor*), incense cedar (*Calocedrus decurrens*), black oak (*Quercus kelloggii*), and interior live oak (*Quercus wislizeni*). The most common tree species in the subalpine zone are whitebark pine (*P. albicaulis*), lodgepole pine (*P. contorta subs. murrayana*), mountain hemlock (*Tsuga mertensiana*), western white pine (*P. monticola*), red fir (*Abies magnifica*), Sierra juniper (*Juniperus occidentalis var. australis*), Jeffrey pine (*P. jeffreyi*), and white fir [32].

Average annual precipitation in the central Sierra ranges from 80 cm at lower elevations to more than 170 cm at higher elevations, with most precipitation falling during the winter as snow [33]. Peak annual SWE levels range from 50 cm to 160 cm, depending on elevation and topography [34].

METHODS

Satellite Image Processing

Near cloud-free imagery from the Landsat sensor was selected from the U. S. Geological Survey Earth Explorer data portal (http://earthexplorer.usgs.gov/) for every year from 1985 to 2015, with the exception of 2012 that was missed during the transition from Landsat 5 to Landsat 8 satellites. Image data from Landsat path/rows 41/35 and 42/34 were acquired between July 15 and August 30 each year, around the peak of the Sierra summer growing season [12,34]. All images used in this study were geometrically registered (UTM Zone 10) using terrain correction algorithms (Level 1T) applied by the U. S. Geological Survey EROS Data Center [35].

For the Landsat 4-5 Thematic Mapper (TM) images acquired at 30-m resolution, surface reflectance data were generated from the Landsat Ecosystem Disturbance Adaptive Processing System [36]. The Moderate Resolution Imaging Spectroradiometer (MODIS) atmospheric correction routines were applied to Level-1 TM data products. Water vapour, ozone, geopotential height, aerosol optical thickness, and digital elevation are included as inputs along with the Landsat data for the Second Simulation of a Satellite Signal in the Solar Spectrum (6S) radiative transfer models to generate top of atmosphere (TOA) reflectance, surface reflectance, brightness temperature, and masks for clouds, cloud shadows, adjacent clouds, land, snow, ice, and water. After 2012, Landsat 8 surface reflectance products were generated from the L8SR algorithm

[36], a method that uses the scene centre for the sun angle calculation and then hard-codes the view zenith angle to 0. The solar zenith and view zenith angles are used for calculations as part of the atmospheric correction.

NDVI calculation

Cloud- and snow-filtered NDVI images were generated as the differential reflectance between the red and near-infrared (NIR) portions of the spectrum by the equation:

NDVI = (NIR – Red) / (NIR + Red)

and scaled from 0 to 10000, where NIR and Red are the reflectance (μm) of the near-infrared and red bands of the electromagnetic spectrum, respectively. Low values of NDVI (near 0) indicate barren land cover whereas high values of NDVI (near 8000) indicate dense canopy vegetation cover. Van Wagtendonk and Root [12] reported that July-August NDVI values greater than 5000 corresponded to high densities of conifer forest and shrubland (chaparral) cover of Yosemite National Park in the central Sierra Nevada. Schiffman [15] found that summer NDVI values greater than 5000 corresponded to an LAI above 1 and NDVI values greater than 7000 corresponded to an LAI above 3 in forests of Yosemite National Park.

NDMI calculation

The severity of drought stress across the study region was determined using the Landsat NDMI [21], computed by the equation:

NDMI = (NIR – SWIR) / (NIR + SWIR)

where, SWIR is the short wave infrared (SWIR; 1.55 – 1.75 μm) band. NDMI (scaled 0 to 10000) is comparable to the Landsat normalized burn ratio (NBR), used for wildfire severity mapping [37]. The reduction of reflectance of the SWIR as compared to the NIR is due to the absorption of water in leaf tissues [38]. High NDMI values represent relatively high vegetation canopy moisture and lower drought stress, while near-zero values would represent relatively low vegetation canopy moisture content and higher drought stress [26].

Other data sets used

Elevation, slope, and aspect at 1 arc-second resolution were determined from the United States Geological Survey (USGS) National Elevation Dataset (NED). Wildfire boundaries and years since fire (YSF) were compiled from the California Department of Forestry, Fire and Resource Assessment Program (FRAP; data available at http://frap.cdf.ca.gov/), with additions from the National Park Service.

DATA ANALYSIS

Yearly NDVI values from 1985 to 2015, at approximately 2020 USFS aerial survey locations (depending on seasonal cloud cover) within Landsat path/row 41/35 and 42/34 areas (Figure 1) were averaged and subjected to time series non-linear (second-order polynomial) regression analysis.

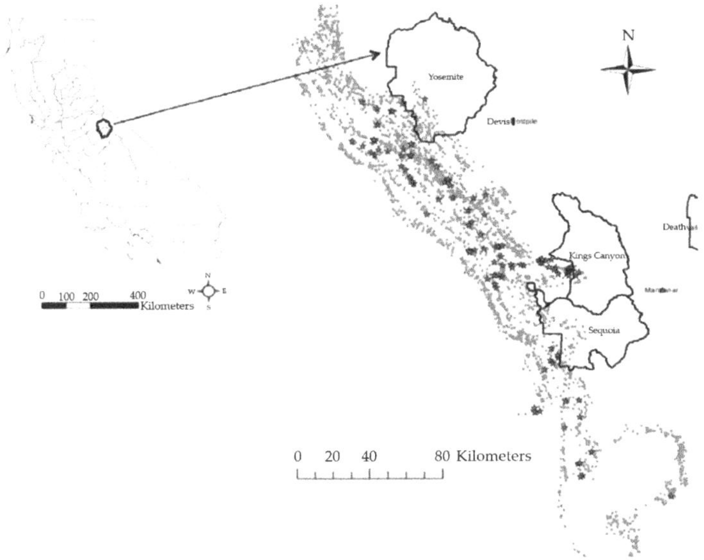

Figure 1: Map of USFS Aerial Survey Points with high tree mortality of the south Sierra Nevada from April, 2015 shaded in grey, and for Landsat dNDVI (2015-2011) < -2000 in star symbols (N = 91).

A significant difference from zero, or between two (yearly) sample means of NDVI or NDMI at a 95% confidence interval ($p < 0.05$), was based on identifying a separation of the mean(s) at greater than (plus or minus) 2 standard errors [39]. Correlations between NDVI (or NDMI) change values at these aerial survey locations with observed tree mortality were also computed, versus latitude, elevation, slope, aspect, and YSF.

RESULTS

NDVI and NDMI time series analysis

For yearly NDVI averaged across all USFS 2015 aerial survey locations with high observed tree mortality, variations over 1985-2015 were dominated by three major periods of extremely low levels of SWE for the Sierra Nevada region [29,34,40]: 1988-1990, 2000-2001, and 2013-2015 (Figure 2). Yearly NDVI declined by 500-700 units (about 10%) during each of these three drought periods. Between each of these drought periods, average NDVI for all April 2015 aerial survey locations recovered rapidly to average pre-drought levels and then increased to exceed those levels by at least 500 NDVI units, or approximately 10% of 1985 averaged NDVI levels.

With focus on the most recent drought period, average NDVI in August 2015 for all USFS aerial survey locations with observed tree mortality in April 2015 was significantly lower than average NDVI in 2011 (pre-drought) or in 2013 ($p < 0.05$). Despite this significant decline in average NDVI from 2011 to 2015, the average NDVI in 2015 was higher than average NDVI in 1990 or in 2002; the years at the ends of the previous two periods of extreme drought in the Sierra Nevada. Nonetheless, the Landsat record showed a major decline of 18% in average NDMI between 2011 and 2015 (Figure 2) for the USDA aerial survey locations. This was statistically significant ($p < 0.05$) and greater than the change in average NDVI for the same set of April 2015 aerial survey locations. The apparent drought stress detected by the steep downward change in NDMI began in 2013 and proceeded at equal downward intervals in both 2014 and 2015. Average NDMI in 2015 for all USFS aerial survey locations with observed tree mortality in April 2015 was significantly lower ($p < 0.05$) than average NDMI in other years of the study period, with the exception of the years 1989 – 1992. Similarly, the negative change in NDVI between 2011 and 2015 (Figure 3) for the aerial survey locations that exceeded a threshold of more than 20% (< -2000 NDVI units; N = 89) showed a similar overall trend to the 30-yr time series of NDMI.

Figure 2: Mean growing season (a) NDVI and (b) NDMI for USFS Aerial Survey Points with high tree mortality from April, 2015 (average N per year, NDVI = 2021 Standard deviation = 99; NDMI =1440 Standard deviation = 177).

Average NDVI Variations with Topography

Slope, aspect, and YSF at the USDA aerial survey locations with high observed tree mortality in April 2015 were each correlated separately using non-linear regression analyses with the Landsat image differences (dNDVI or dNDMI) between 2015 and 2011. None of these three geographic variables explained a significant portion of the variation among the sampled 2015-2011 dNDVI or dNDMI values. The resulting R^2 correlation values for dNDVI and dNDMI were lower than 0.05 in the regression analysis results. It is worth noting however that the majority of aerial survey locations with the most extreme negative 2015-2011 dNDVI values (in excess of -2000 units) were detected at slopes between 15% and 30%.

Correlations with latitude of sampled 2015-2011 dNDVI or dNDMI values at the USDA aerial survey locations with observed tree mortality in April 2015 were not significant, although the most extreme negative 2015-2011 dNDVI and dNDMI values were located mainly between 36° and 37° N latitude (Figure 4). The correlation with elevation of sampled 2015-2011 dNDMI values at the USDA aerial survey locations was significant ($p < 0.05$) and showed that most of the extreme negative 2015-2011 dNDMI values in excess of -2000 units were detected at elevations below 2000 m. The correlation with elevation for the most extreme negative 2015-2011 dNDVI values was also significant ($p < 0.05$; Figure 5), but in contrast to dNDMI, showed a decreasing pattern of dNDVI with increasing elevation, starting at 1200 m to approximately 2300 m elevation. This may be due, in part, to the predominately lower elevation (< 2100) survey coverage of the April 2015 USDA flights; whereas surveys later in the summer detected extensive tree mortality at higher elevations in the southern Sierra.

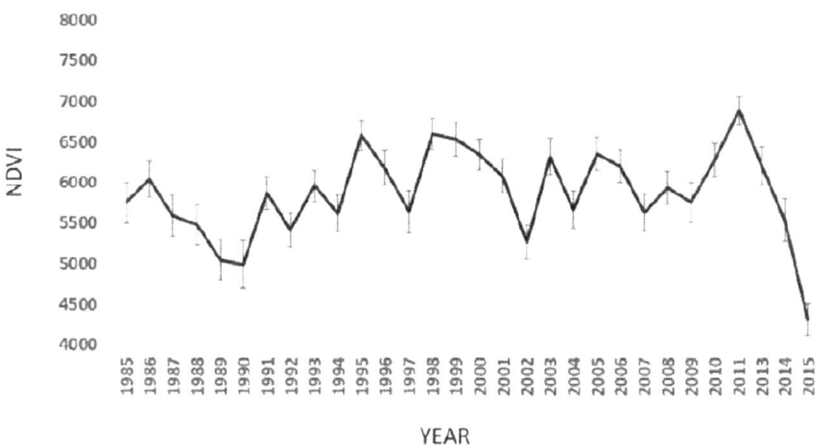

Figure 3: Mean growing season NDVI for the subset of USFS aerial survey points with high tree mortality from April 2015 with dNDVI (2015-2011) < -2000 (average N per year = 89, Standard Deviation = 5).

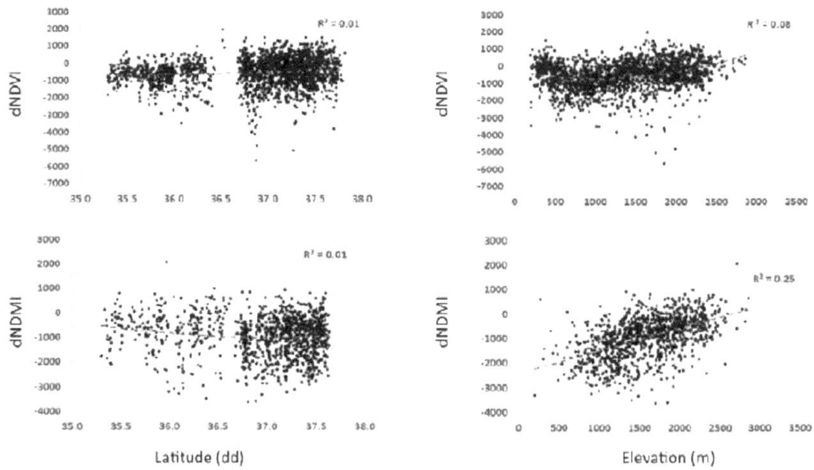

Figure 4: Correlations of 2015-2011 growing season dNDVI and dNDMI with latitude and elevation for USFS Aerial Survey Points with high tree mortality from April, 2015 (N points, NDVI = 1945, NDMI = 1064).

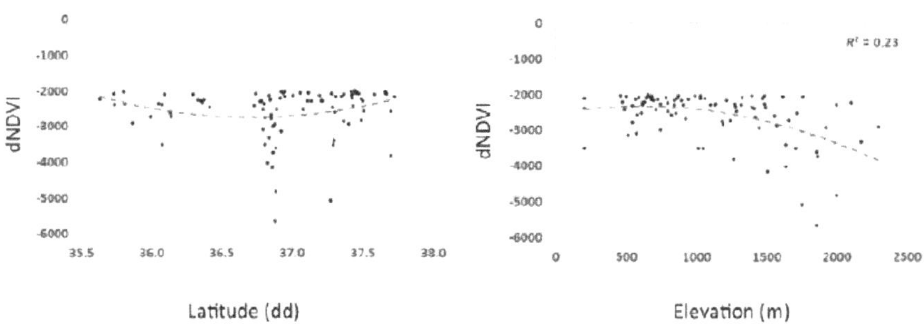

Figure 5: Correlation of latitude and elevation with growing season dNDVI for the subset of USFS aerial survey points with high tree mortality from (April 2015) with dNDVI (2015-2011) <- 2000 (N = 91).

Locations of greatest negative NDVI change

A regional scale view of the 2015-2011 dNDVI values mapped from Landsat imagery over the study area (Figure 6) showed extensive areas of negative dNDVI (in excess of -2000 units) on the southwestern margins of Yosemite NP, Kings Canyon NP, and Sequoia NP, and the Sierra and Sequoia NF boundaries (Figure 6). For comparison of these apparent drought impacts on tree die-back, the areas within five large wildfires that occurred across the region in the period of 2013 to 2015 showed 2015-2011 negative dNDVI values in excess of -3000 units. The majority of USDA aerial survey locations with April 2015 tree mortality and negative 2015-2011 dNDVI values in excess of -3000 units were within the July 2015 Rough Fire perimeter, which started on the western boundary of Kings Canyon NP and burned into Cedar Grove during September 2015.

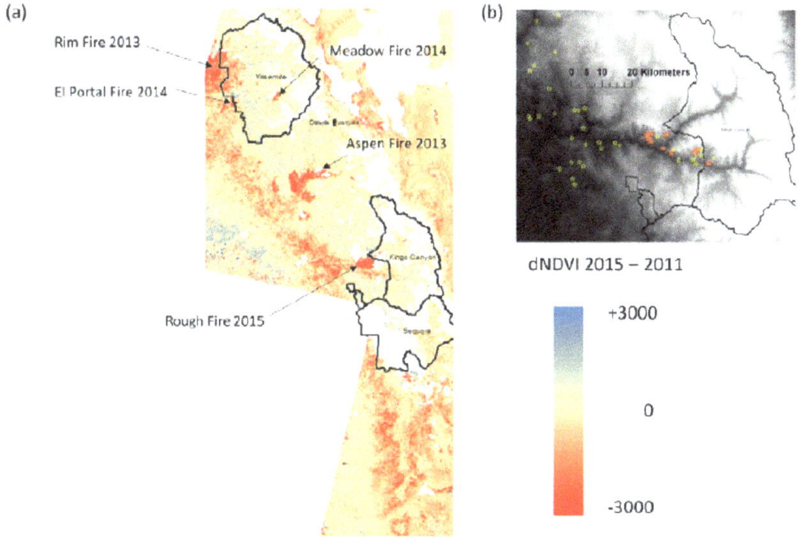

Figure 6: (a) Map of Landsat dNDVI for 2015-2011 for the southern Sierra Nevada foothill and mountain region; (b) Map of USFS Aerial Survey Points with high tree mortality from April, 2015 near Kings Canyon NP in yellow star symbols; with dNDVI of 2015-2011 < -3500 in red star symbols.

DISCUSSION

Results from this study indicate that changes in Landsat drought sensitive indices from 2011 to 2015 matched patterns of tree mortality across USFS April 2015 aerial survey locations in the southern Sierra Nevada. Both NDVI and NDMI had been increasing steadily since the early 1990s across these aerial survey locations, which was indicative of infilling by relatively young forest stands with high stem densities [41,42]. Drought and the historically low snow year of 2015 may have essentially reset the average forest canopy density (per LAI and water use) for many forests in the region to late 1980s levels, due to drought related mortality and stand removal by large wildfires.

In support of these conclusions, evidence strongly linking satellite NDVI and NDMI to forest growth and decline patterns comes from a series of previous ground validation studies in the Sierra Nevada. For instance, forest stand NDVI values greater than 5000 have been correlated with dense foliar cover in conifer forest stands in the Sierra (van Wagtendonk and Root, 2003). Schiffman (2008) also reported a close correlation between NDVI and leaf area index (LAI) measured across Sierra forest stands. These and other ground-truth studies [17,18] have established NDVI as a reliable remotely sensed surrogate for variations in stand LAI and canopy biomass in this montane region.

Several field measurement data sets of conifer forest growth and stand structure in the western U. S. have repeatedly demonstrated a strong linear relationship between LAI and stem basal area growth. For example, Kashian et al. [43] showed a significant relationship ($R^2 = 0.6$, $p < 0.01$) between LAI and basal area increment (BAI) for 48 lodgepole pine stands aged 45-350 years in Yellowstone National Park, Wyoming. Stem density at the stand level was weakly correlated with both LAI and BAI ($R^2 = 0.22$) overall and no correlation was observed with either LAI or BAI at stand age classes less than 200 years old. Previous measurements by Knight et al. [44] at eight stands of lodgepole pine in the Medicine Bow Mountains of south-eastern Wyoming showed the same basic patterns between LAI, basal area, and stem density. Peterson et al. [45] reported a significant correlation ($R^2 = 0.9$, $p < 0.01$) between LAI and BA across six lodgepole and whitebark pine stands in the Eastern Brook Lakes watershed of the Inyo National Forest in the Sierra Nevada. There was no correlation here with stem density at the stand level and LAI or BA ($R^2 = 0.1$). Based on this review of numerous field plot measurements, LAI and (hence NDVI) should be a reliable predictor of stem growth in Sierra Nevada forest stands.

Ceccato et al. [46] developed and tested an index similar to the NDMI using data from the SPOT satellite sensor. They concluded that the SWIR channel was critical to estimating leaf water content and that the NIR channel was needed to account for variation of leaf internal structure and dry matter content variations. Likewise, Van Gunst et al. [27] used multi-year Landsat NDMI data to examine variations in mortality among forest types and climate periods from 1985 to 2010 in the mixed conifer forest of the Lake Tahoe Basin. Results showed that low Landsat NDMI values (< 2000) were significantly related to low forest canopy vigor and high annual tree mortality rates of > 20%. This study also found that positive density-dependent mortality was associated with lower elevation (below 2100 m) forests and drier climate periods. In mid- to upper-elevation (above 2100 m) forests, increased density was more often associated with decreased probability of mortality, especially during wetter periods. Tree mortality was highest during the drought period of 1988 to 1994, but not subsequent drought periods (prior to 2011) of similar severity. Results also suggested increased risk of tree mortality on north-facing slopes across all forests and all climate periods.

Several previous field studies have shed light on the mechanisms of forest decline the Sierra Nevada range. For instance, Smith et al. [47] surveyed trees in the southern Sierra National Forest and reported that pathogen- and insect-associated mortality was significantly greater in areas of high stand density but was not higher for shade-tolerant species. Tree mortality was higher than expected for large-diameter trees, suggesting an acceleration of old- tree mortality under current fire suppression conditions. The overall conclusion of this study was that the increase in stem density due to fire suppression and/or climate change has overwhelmed the corrective nature of mortality by native pathogens and insects in the forest of the southern Sierra Nevada range, at least up until the 2013-2015 droughts. Related to this, Dolanc et al. [41] speculated that the decrease in density of large trees in the Sierra Nevada subalpine zone over the past 75 years has been consistent with observed changes in climate. Higher temperatures may have shortened the time to drought induced mortality of trees and can increase the susceptibility of conifers to insect attacks [33,48]. Guarin and Taylor [49] investigated the influence of drought and topography on patterns of tree mortality in old-growth mixed conifer forests in Yosemite NP. Correlations between tree mortality and drought were evident only for multi-year periods of 2 to 5 years and the frequency of tree death dates was negatively correlated with April snowpack depth recorded since the mid-1970s. The density of dead trees was found to be slightly higher on north- than south-facing slopes, possibly due to competition between white fir and incense cedar seedlings and saplings.

As a comparison to these latest drought study results from the southern Sierra Nevada region, Potter [26] used both Landsat dNDVI and dNDMI to assess the impacts of the 2013-2014 drought on the forested ecosystems of the California Central Coast. In contrast to the extensive drought stress detected in the southern Sierra Nevada region (Figure 6), only a small fraction (less than 5%) of all forest cover area in the Central Coast region showed relatively high drought stress. The Central Coast is not snow-dependent in terms of the annual water supply for woody vegetation communities (as is the Sierra Nevada), which, together with a relatively short return interval promoting rapid regrowth between stand replacing wildfires on the coast, may explain the observed lack of a severe drought impact since 2010-11 on forests in this ecosystem.

From a general remote sensing perspective, imaging by hyperspectral sensors (using a relatively narrow band resolution of several nm) can play a role in drought impact monitoring [50], provided that aircraft flights are conducted in a timely, economical, and sustained manner. Findings from tropical forest experiments with hyperspectral imaging have shown that nearly 20% of canopy LAI can be lost from the wet to dry season during enhanced drought conditions [51,52]. Leaf water potential was also highly sensitive to drought, decreasing by almost 30% compared to the control forest. The net effect on remotely sensed hyperspectral data was a measurable lowering of reflectance across the entire NIR spectral range in drought affected areas, compared with that of unaffected control plots. Similarly, in a hyperspectral sensing study of California oak forest response to moisture stress, Pu et al. [53] reported the same kind of broad, uniform lowering of signal across the entire NIR reflectance spectra. In theory, hyperspectral imaging is better suited to detecting relatively narrow band responses (of several nm) to changing environmental factors. Both Landsat NDVI and NDMI adequately account for the spectrally uniform, broad-band NIR response to drought stress (as commonly seen in hyperspectral imagery), while meeting all the criteria of being timely, economical, with sustained observations spanning over 30 years across all forest ecosystems. The main compromise with Landsat satellite versus airborne imaging is in the spatial resolution of the imagery, from a few meters to 30 m ground resolution.

CONCLUSIONS

According to Landsat image mapping, the regional patterns of forest disturbance and potential canopy dieback over the period 2011 to 2015 in the southern Sierra Nevada are a result of a combination of historically large wildfires and historically low annual snow accumulation amounts. Both Landsat dNDVI and dNDMI time series were found to closely match tree mortality observations from USFS April 2015 aerial survey locations, and verified previously published findings, which indicated that low Landsat

NDMI values (a change to below 2000 NDMI units) were significantly related to low forest canopy vigor and high annual tree mortality rates of > 20%. Analysis of Landsat image data in subsequent years around these same southern Sierra Nevada aerial survey locations may assist forest managers in the assessment of delayed tree mortality post-2015, and in tracking subsequent regrowth rates of drought-affected forest stands during more favourable precipitation years.

ACKNOWLEDGEMENTS

The author thanks Jeffrey Moore (USFS, Davis, California) for valuable comments on the results of this study.

REFERENCES

1. 1. Anderegg WRL, Flint A, Huang C, Flint L, Berry JA, et al. (2015) Tree mortality predicted from drought-induced vascular damage. Nature Geosci 8: 367-371.
2. 2. Moore J, Heath Z, Bulaon B (2015) Aerial Detection Survey - April 15th - 17th 2015, Forest Health Protection Survey, United States Forest Service, pp.6. pp.
3. 3. Collins B, Woodcock CE (1996) An assessment of several linear change detection techniques for mapping forest mortality using multitemporal Landsat TM data. Remote Sens Environ 56: 66 -77.
4. 4. Amiro BD, Chen JM, Liu J (2000) Net primary productivity following forest fire for Canadian ecoregions. Can J For Res 30: 939-947.
5. 5. Rogan J, Franklin J (2001) Mapping wildfire burn severity in southern California forests and shrub lands using Enhanced Thematic Mapper imagery. Geocarto Int 16: 89- 99.
6. 6. Rogan J, Miller J, Stow DA, Franklin J, Levien L, et al. (2003) Land-cover change monitoring with classification trees Using Landsat TM and ancillary data. Photogramm Eng Remote Sens 69: 793-804.
7. 7. Fischer L, Rosenberg M, Mahon L, Liu Z, Maurizi B, et al. (2004) Monitoring land cover changes in California, a USFS and CDF cooperative program, Northern Sierra Project Area - Cycle II. State of California, Resources Agency, Department of Forestry and Fire Protection, Sacramento, CA.
8. 8. Epting J, Verbyla DL (2005) Landscape level interactions of pre-fire vegetation, burn severity, and post-fire vegetation over a 16-year period in interior Alaska. Can For Res 35: 1367-1377.
9. 9. Cuevas-Gonzalez M, Gerard F, Balzter H, Riano D (2009) Analysing forest recovery after wildfire disturbance in boreal Siberia using remotely sensed vegetation indices. Global Change Biol 15: 561-577.
10. 10. Casady GM, Marsh SE, (2010) Broad-scale environmental conditions responsible for post-fire vegetation dynamics. Remote Sens 2: 2643-2664.
11. 11. Li S, Potter C (2012) Vegetation regrowth trends in post forest fire ecosystems across North America from 2000 to 2010. Nat Sci 4: 755-770.

12. 12. van Wagtendonk JW, Root RR (2003) The use of multi-temporal Landsat Normalized Difference Vegetation Index (NDVI) data for mapping fuel models in Yosemite National Park, USA. Int J Remote Sens 24: 1639-1651.

13. 13. Potter C, Li S, Huang S, Crabtree RL (2012) Analysis of sapling density regeneration in Yellowstone National Park with hyperspectral remote sensing data. Remote Sens Environ 121: 61-68.

14. 14. Franks S, Masek JG, Turner MG (2013) Monitoring forest regrowth following large scale fire using satellite data - A case study of Yellowstone National Park, USA. Eur J Remote Sens 46: 561-569.

15. 15. Schiffman B (2008) Estimation of Leaf Area Index (LAI) through the acquisition of ground truth data in Yosemite National Park, ASPRS 2008. Annual Conference Portland, Oregon.

16. 16. Chen JM, Cihlar J (1996) Retrieving leaf area index of boreal conifer forests using Landsat TM images. Remote Sens Environ 55: 153-162.

17. 17. Gong P, Pu R, Miller JR (1995) Coniferous forest leaf area index estimation along the Oregon transect using compact airborne spectrographic imager data. Photogramm Eng Remote Sens 61: 1107-1117.

18. 18. Turner DP, Cohen WB, Kennedy RE, Fassnacht KS, Briggs JM (1999) Relationships between leaf area index and Landsat TM spectral vegetation indices across three temperate zone sites. Remote Sens Environ 70: 52-68.

19. 19. Wang J, Rich PM, Price K, Kettle W (2004) Relations between NDVI and tree productivity in the central Great Plains. Int J Remote Sens 25: 3127-3138.

20. 20. Li S, Potter C, Hiatt C (2012) Monitoring of net primary production in California rangelands using Landsat and MODIS satellite remote sensing. Nat Res 3: 56-65.

21. 21. Gao BC (1996) NDWI - A normalized difference water index for remote sensing of vegetation liquid water from space. Remote Sens Environ 58: 257-266.

22. 22. Hunt ER, Rock BN (1989) Detection of changes in leaf water content using near and middle-infrared reflectances. Remote Sens Environ 30: 43-54.

23. 23. Chuvieco E, Riano D, Aguado I, Cocero D (2002) Estimation of fuel moisture content from multitemporal analysis of Landsat Thematic Mapper reflectance data: application in fire danger assessment. Int J Remote Sens 23: 2145-2162.

24. 24. Dennison PE, Roberts DA, Peterson SH, Rachel J (2005) Use of normalized difference water index for monitoring live fuel moisture. Int J Remote Sen 26: 1035-1042.

25. 25. Goodwin NR, Coops NC, Wulder MA, Gillanders S, Schroeder TA, et al. (2008) Estimation of insect dynamics using a temporal sequence of Landsat data. Remote Sens Environ 112: 3680-3689.

26. 26. Potter CS (2015) Assessment of the immediate impacts of the 2013-2014 drought on ecosystems of the California central coast. West N Am Naturalist 75: 129-145.

27. 27. Van Gunst KJV, Weisberg P J, Yang J, Fan Y (2015) Do denser forests have greater risk of tree mortality: A remote sensing analysis of density-dependent forest mortality. Forest Ecol Manag 359: 19–32.

28. 28. Hanak E, Mount J, Chappelle C (2014) California's Latest Drought, Public Policy, Institute of California, February 2014.

29. 29. Belmecheri S, Babst F, Wahl E, Stahle D, Trouet V (2015) Multi-century evaluation of Sierra Nevada snowpack, Nature Clim Change 6: 2-3.

30. 30. Miller JD, Safford H (2012) Trends in wildfire severity: 1984 to 2010 in the Sierra evada, Modoc Plateau, and southern Cascades, California, USA. Fire Ecol 8: 41-57.

31. 31. Safford HD, Van de Water K, Schmidt D (2011) California Fire Return Interval Departure (FRID) map, 2010 version. USDA Forest Service, Pacific Southwest Region and The Nature Conservancy-California.

32. 32. Rundel P, Kaufman J, Nathan S, Weixelman D (2007) Montane and subalpine vegetation of the Sierra Nevada and Cascade Ranges. In: Barbour, MG Major (3rdedn.). J Terrestrial vegetation of California. New York, USA, pp: 559–599.

33. 33. Lutz JA, Wagtendonk JWV, Franklin JF (2009) Twentieth-century decline of large diameter trees in Yosemite National Park, California, USA. Forest Ecol Manag 257: 2296-2307.

34. 34. Trujillo E, Molotch NP, Goulden ML, Kelly AE, Bales RC (2012) Elevation dependent influence of snow accumulation on forest greening. Nat Geosci 5: 705-709.

35. 35. Chander G, Markham B, Helder D (2009) Summary of current radiometric calibration coefficients for Landsat MSS, TM, ETM+, and EO-1 ALI sensors. Remote Sens Environ 113: 893-903.

36. 36. Masek JG, Vermote EF, Saleous N, Wolfe R, Hall FG, et al. (2006) A Landsat surface reflectance data set for North America, 1990-2000. GRSL 3: 68-72.

37. 37. Miller JD, Yool SR (2002) Mapping forest post-fire canopy consumption in several overstory types using multi-temporal Landsat TM and ETM data. Remote Sens Environment 82: 481-496.

38. 38. Lozano FJ, Suárez-Seoane S, De Luis E (2007) Assessment of several spectral indices derived from multi-temporal Landsat data for fire occurrence probability modeling. Remote Sens Environ 107: 533-544.

39. 39. Gelman A, Hill J (2007) Data analysis using regression and multilevel/hierarchical models. Cambridge University Press, New York, New York, USA.

40. 40. Kapnick S, Hall A (2010) Observed climate-snowpack relationships in California and their implications for the future. J Clim 23: 3446-3456.

41. 41. Dolanc CR, Thorne JH, Safford HD (2013) Widespread shifts in the demographic structure of subalpine forests in the Sierra Nevada, California, 1934 to 2007. Global Ecol Biogeogr 22: 264-276.

42. 42. McIntyre PJ, Thorne JH, Dolanc CR, Flint AL, Flint LE, et al. (2015) Twentieth-century shifts in forest structure in California: Denser forests, smaller

trees, and increased dominance of oaks. Proceedings of the National Academy of Sciences 112: 1458-1463.

43. 43. Kashian DM, Turner MG, Romme WH (2005) Variability in leaf area and stemwood increment along a 300-year lodgepole pine chronosequence. Ecosystems 8: 48-61.
44. 44. Knight DH, Fahey TJ, Running SW (1985) Water and nutrient outflow from contrasting lodgepole pine forests in Wyoming. Ecol Model 55: 29-48.
45. 45. Peterson DL, Arbaugh MJ, Lardmer MA (1989) Leaf area of lodgepole pine and whitebark pine in a subalpine Sierra Nevada forest. Can J For Res 19: 401-403.
46. 46. Ceccato P, Flasse S, Gregoire, J. (2002) Designing a spectral index to estimate vegetation water content from remote sensing data: Part 2. Validation and applications. Remote Sens Environ 82: 198-207.
47. 47. Smith TF, Rizzo DM, North M (2005) Patterns of mortality in an old-growth mixed- conifer forest of the southern Sierra Nevada, California. Forest Science 51: 266-275.
48. 48. Allen CD, Macalady AK, Chenchouni H, Bachelet D, McDowell N, et al. (2010) A global overview of drought and heat-induced tree mortality reveals emerging climate change risk for forests. Forest Ecol Manag 259: 660-684.
49. 49. Guarin A, Taylor AH (2005) Drought triggered tree mortality in mixed conifer forests in Yosemite National Park, California, USA. Forest Ecol Manag 218: 229-244.
50. 50. Asner GP, Nepstad D, Cardinot G, Ray D (2004) Drought stress and carbon uptake in an Amazon forest measured with spaceborne imaging spectroscopy. Proc Natl Acad Sci USA 101: 6039-6044.
51. 51. Nepstad DC, Moutinho P, Dias MB, Davidson E, Cardinot G, et al. (2002) The effects of partial throughfall exclusion on canopy processes, aboveground production, and biogeochemistry of an Amazon forest. J Geophys Res 107: 1-18.
52. 52. Asner GP, Alencar A (2010) Drought impacts on the Amazon forest: the remote sensing perspective. New Phytologist 187: 569-578.
53. 53. Pu R, Kelly M, Anderson GL, Gong P (2008) Using CASI hyperspectral imagery to detect mortality and vegetation stress associated with a new hardwood forest disease. Photogramm Eng Remote Sens 74: 65-75.

15
CHAPTER

PHOTOGRAMMETRIC METHODOLOGY FOR THE PRODUCTION OF GEOMORPHOLOGIC MAPS: APPLICATION TO THE VELETA ROCK GLACIER (SIERRA NEVADA, GRANADA, SPAIN)

Javier de Matías, José Juan de Sanjosé, Gonzalo López-Nicolás, Carlos Sagüés and José Jesús Guerrero

Keywords: Sierra Nevada (Spain); rock glacier; photogrammetry; computer vision; matching

INTRODUCTION

The study of rock glaciers in high-mountain zones requires the production of detailed maps for the analysis of their evolution and behavior through time. Using geomatic techniques (geodesy, global positioning system, photogrammetry) it is possible to study the dynamics of geomorphologic structures with high precision. In order to obtain 3D reconstructions in rock glaciers, conventional photogrammetry (metric camera and normal shots) and cheaper methods (semi-metric cameras and convergent photographs) can be used [1,2,3,4], but we aim for a more automatic method.

Obtaining detailed maps (cartography) in rock glaciers is difficult, because the use of flights to obtain photogrammetric images in these areas is expensive. Our aim is to develop a quick and cheap method to obtain enough 3D points to create a detailed DTM. Thus, two methodological goals are considered:

- Cartographic representation of the rock glacier, for example at a scale of 1/1,000 with a margin of error in the points of less than ±20 cm. This is the main objective of this study.
- Determination of the glacier dynamics through the time. Currently, we are working in this new methodology.

Citation: Javier de Matías, José Juan de Sanjosé, Gonzalo López-Nicolás, Carlos Sagüés and José Jesús Guerrero, Photogrammetric methodology for the production of geomorphologic maps: application to the veleta rock glacier (Sierra Nevada, Granada, Spain), doi:10.3390/rs1040829

To fulfill these goals we use computer vision techniques to automatically detect as many points as possible of the natural environment which is specific to our example Veleta rock glacier. The application of computer vision techniques in this field has the following objectives:

- The creation of a Digital Terrain Model (DTM) with enough points in order to obtain a geomorphic map of the Veleta rock glacier.
- Replacing costly photogrammetric flights using planes with less expensive photographs taken from the top of the surrounding mountains, or using balloons or hang-gliders.
- Elimination of extensive office work by a human operator in obtaining 3D points or contour lines by replacing photogrammetric restitution with an automatic method, which also eliminates human errors.

CHARACTERISTIC OF THE VELETA ROCK GLACIER

The Veleta rock glacier is located in "Sierra Nevada" (37°N–3°W) in a cirque at the bottom of Mount "Veleta" (3,398 m. a.s.l.). Sierra Nevada is a National Park at the south of Spain (Figure 1a). This mountain is very vulnerable to gelifraction which results in much more rock material on the Veleta rock glacier. The Veleta rock glacier forms an open figure "╗" (Figure 1b). Its source is attached to the wall of the "Corral" and continues in a northern direction and later extends to the west direction. It has an average slope of 20°, with the front at 3,090 m. a.s.l. and the root at 3,175 m. a.s.l. It is a tongue-shaped rock glacier 35 m wide and 109 m long [5].

The Veleta rock glacier has horizontal and vertical displacement with volumetric changes which have been verified through geodetic measurements over an eight year period (2001–2008). During this period the average dynamic of the rock glacier was 0.40/0.60 m in planimetry and −2.10/−2.60 m in altimetry [1].

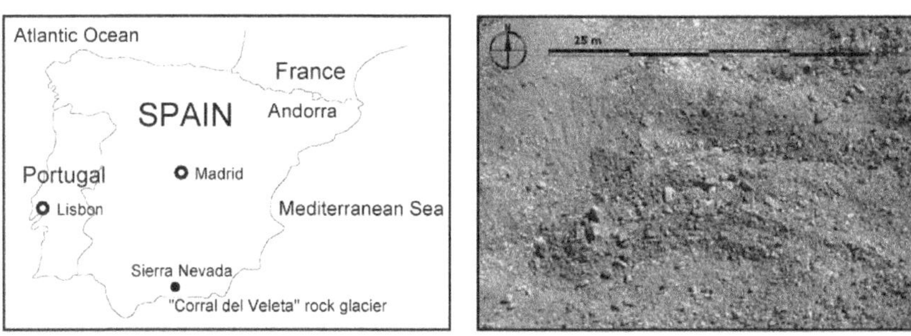

(a) (b)

Figure 1: (a). Veleta rock glacier situation; (b). Close view of the glacier from Mount Veleta.

GEOMATIC METHODS

Several geomatic techniques can be employed to study any kind of rock glacier: Interferometry radar, laser scanner, geodesy (angle and distance measurements), levelling (geometric, trigonometric), Global Positioning System (GPS), photogrammetry (aerial, terrestrial, close range) [6,7,8]. In the case of Veleta rock glacier, the geodetic techniques (high precision total station), GPS and close range photogrammetry have been used [1]. The orography around the Veleta rock glacier (300 m high walls) and the cost make the application of some geomatic techniques difficult.

Some authors have used geodetic techniques and aerial photogrammety to obtain maps or to measure the glacier dynamics [3,9,10,11]. The advantage of the photogrammetry (aerial, terrestrial) over the geodesy is that photogrammetry can show the position of many points (photographic information), and it makes collecting information in the field much faster than geodesy [12,13].

Geodetic Survey and GPS

On the surface of the Veleta rock glacier, 20 control points were installed to monitor the glacier dynamics. These control points (iron rods) are measured every year to observe glacier dynamics and are also used for the photogrammetric process.

High precision topographic measurements were performed using a total station and a GPS. The total station is placed on the stable lateral moraine of the Veleta rock glacier and the maximum distance between the survey stations and the control points is 180 meters. The GPS method used is "RTK static". So, a receiver is fixed on a point of known coordinates (X,Y,Z) and the other receiver is successively placed at the iron rods for 1 minute to ensure that every point has a spatial precision or GDOP less than 3 cm [14]. The control points for photogrammetry are the same points used for the monitoring of the dynamics of the rock glacier in order to reduce the time of data collection (Figure 2a).

(a)

(b)

Figure 2: (a) Photogrammetric control point; (b) Photogrametric shots from mount Veleta towards the rock glacier.

Close Range Photogrammetry (Conventional Photogrammetry)

A calibrated camera should be used to apply a photogrammetric general method and then the principal point, the focal length, the grid reseau coordinates and the lens distortion are known.

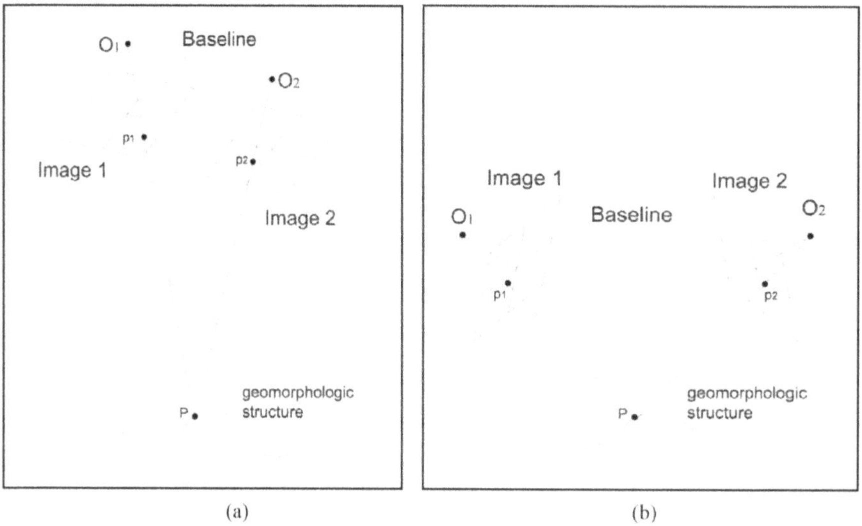

Figure 3: (a) Geometry of normal photogrammetry; (b) Geometry of convergent images. O_1 and O_2 are the centres of projection of the photographs.

In the general aerial photogrammetric method, the cameras are placed in the normal configuration in a vertical direction in such a way that the optical axes of both photographs are parallel and perpendicular to the baseline (Figure 3a). This can be done with a plane flight but it cannot be done by hand in the Veleta rock glacier. Here, the operator is on the Veleta peak with the camera in hand on an uneven surface (Figure 2b). The images obtained have a convergent geometry (Figure 3b), and we take advantage of the epipolar geometry concept to deal with these convergent photographs. We have taken several photographs from which we have selected pairs to carry out the photogrammetric process using a stereoplotter.

The photogrammetric method of restitution to obtain the internal, relative and absolute orientations was difficult and it took several trials [15,16]. This is because the traditional restitution method is designed to work with image pairs in normal or quasi-normal configuration with parallel optical axis.

In summary, the general photogrammetric method used to obtain the map of the Veleta rock glacier involved the following issues:

- Method of photogrammetry with analytical plotter "Leica SD-2000".
- The final result of accuracy in orientation was 6 cm planimetry (X, Y) and 8 cm altimetry (Z).
- The map was made from the stereoscopic orientation of the image pair through contour lines with the PRO600 software (Figure 4).

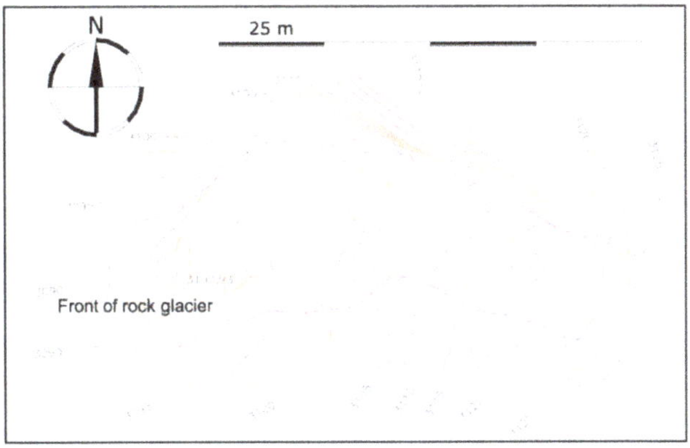

Figure 4: Map of the Veleta rock glacier (scale: 1/1,000, contour lines: 1 m).

COMPUTER VISION FOR PHOTOGRAMMETRIC RECONSTRUCTION

The goal of the photogrammetric method presented in this work is to obtain automatically a 3D model of a geomorphologic structure. This method develops preliminary work presented in the Sixth International Conference on Geomorphology [17]. The methodology proposed is supported by means of techniques developed in the field of Computer Vision. The input requirements of the method are: a set of images of the geomorphologic structure taken from different points of view and the camera calibration (focal length, principal point and radial distortion). For the Veleta rock glacier example we work using a digital camera (Canon EOS 5D with 35 mm lens). In this section we present the different method steps which are summarized as follows:

- Extraction features: Extraction of point features of each image using the SIFT extractor [17,18].
- Matching of points (between image pairs).
- Robust matching through the epipolar geometry: The epipolar geometry (a geometric constraint) of the image pair is computed by using robust statistic techniques. Matches that do not hold the epipolar constraint are removed as mismatches.
- Camera processing (pose estimation): We know the internal parameters of camera (focal lenght, principal point, radial distortion) and we estimate the rotation and translation of the cameras.
- DTM generation: This step requires the camera projection matrices and a triangulation process to obtain a 3D model of the geomorphologic structure.

Extraction of Features

In the first step, feature points are extracted from each image. The more usual image features are points [19], or lines [20]. Here we choose point features because they are more easily extracted in nonstructured environments. We use DoG (Difference of Gaussian) detector and SIFT descriptor to extract a set of point features [18].

The regions extracted with DoG detector are described with a vector of dimension 128 and the descriptor vector is divided by the square root of the sum of the squared components to get illumination invariance. The descriptor is a 3D histogram of gradient location and orientation. SIFT descriptors are very interesting because they are invariant to scale and rotation and have good properties with regard to repeatability, distinctiveness, and robustness, and then, SIFT has shown to be very useful for point matching purposes. For example, we can distinguish the same point in several images even with big scale and orientation changes. In this way we obtain a set of point features for each image of stereo pair using SIFT descriptors, which we use for find pair of corresponding points.

Although there are other descriptors or methods it was demonstrated with different measures that the SIFT descriptors are superior to most others. Several works have compared existing descriptors and concluded that SIFT is a good feature point extractor method in a general use or in photogrammetric applications [21].

Matching of Points

The SIFT method provides a set of image locations and descriptors. We will use this information for matching points process. We obtain a list of image 1 descriptors and a list of image 2 descriptors that form a database of keypoints. The best candidate for a matching is his closer keypoint neighbor in database, that is to say, the keypoint with lowest euclidean distance respect the correspondent descriptor. Nevertheless, many of the image points do not have an equivalent matching point. Thus, it is necessary a way to remove points without a good matching in database.

A priori, the more successful solution is to establish a threshold for the distance to the closer descriptor and even better to compare the distance between the closer neighbor and second closer neighbor. Using this method we remove 90% of wrong matched points and we lose only 5% of correct ones [18]. As we can see in the next sections, SIFT matched points are very good for obtaining an accurate DTM of rock glacier area.

Robust Matching through the Epipolar Geometry

In this step the epipolar constraint is exploited to remove the wrong matches which still remain. Now, we briefly introduce some concepts related to the epipolar geometry [22] and how it can be used to perform a robust matching [23].

An image taken by a camera is the projection of a 3D scene in a 2D plane. Then, for each 3D point $\mathbf{P} = (X,Y,Z)$ we have a projected point in the image plane with coordinates $\mathbf{p}_1 = (u_1,v_1,1)$ in the first image and $\mathbf{p}_2 = (u_2,v_2,1)$ in the second. The epipolar geometry represents the intrisic geometry between two views. It is independent of the scene structure and only depends on the relative localization between the cameras (rotation and translation) and the internal camera calibration. The baseline is defined as the straight line that joins the optical centers of the cameras. The epipoles $(\mathbf{e}_1,\mathbf{e}_2)$

are the intersection points of the baseline with the image planes. The fundamental matrix F is the algebraic representation of the epipolar geometry and it satisfies the so called epipolar constraint: $p_2^T F p_1 = 0$. Therefore, each pair of matched points (p_1,p_2) must hold this constraint; otherwise it is a wrong match.

Each correspondence gives one equation through the epipolar constraint. The fundamental matrix is a 3 × 3 matrix and therefore it can be computed up to scale by solving a linear system from a minimum set of 8 points matches [22]. Nevertheless we must assume that outliers and mismatches are present in the set of point matches. Thus, the fundamental matrix is computed iteratively with the RANSAC method [24], which is robust in presence of outliers. The result of this step is the fundamental matrix relating the pair of images. The point matches that support this fundamental matrix are considered good correspondences and the rest of matches are discarded as wrong matches.

Camera Processing (Pose Estimation)

We use a previously calibrated camera (focal length, principal point, radial distortion). Therefore if we want to get the complete camera system we need to calculate the exterior orientation (rotation and translation of cameras or camera pose).

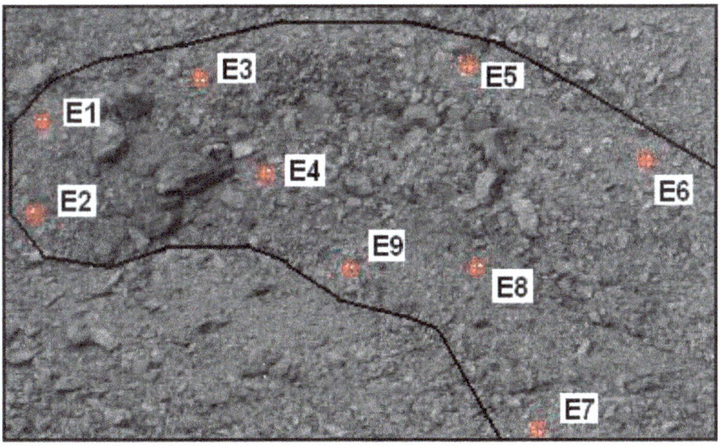

Figure 5: Photograph of the distribution of control points (E1–E9) denoted with marks, around the Veleta rock glacier (2008).

We use a nonlinear method to calculate the exterior orientation: The bundle adjustment [25]. The purpose of bundle adjustment is to get an optimization to minimize the reprojection error through a Gradient Descent solution. We calculate an inicialization using DLT (Direct Linear Transform) and use an iterative approximation to minimize reprojection error. In this bundle adjustment process we use a set of control points, that is to say points with known 3D coordinates (Figure 5). Although bundle adjustment could be used to obtain a self-calibration, we only use it for exterior orientation process (pose camera estimation) since we know camera calibration.

DTM Generation

In this step the 3D reconstruction of the scene is performed to produce the DTM from the set of projection camera matrices. We know the internal camera parameters and the exterior orientation (pose estimation) from the camera processing step in section 4.4, therefore we have now a calibrated system.

As we know the camera projection matrices, the 3D structure may be recovered by triangulation [22]. A simple scheme is proposed using a linear triangulation method to recover the 3D structure [26]. This scheme is based in the fact that the image points have noise and therefore the rays back-projected from the noisy matches in the images are skew in the space. Since the rays do not intersect in general in just one point, the measured points do not satisfy exactly the epipolar constraint. Then, triangulation by back-projecting rays from the matches will fail. This problem is solved if the match points are corrected by minimizing a cost function which represents errors in the image. Once the points are corrected, a linear triangulation method can be carried out to obtain the 3D structure. This 3D structure can be improved using bundle adjustment process as we do for pose estimation. At the end of the triangulation we get a cloud of 3D points. The last step is to convert this 3D point cloud in a surface using a Delaunay triangulation.

RESULTS AND DISCUSSION

We test the proposed methodology in Veleta rock glacier using 2008 campaign images. As we explain in previous sections we obtained these images using a digital camera, specifically a Canon EOS 5D with a 35 mm lens. We can use the proposed methodology with all the possible pair combinations, but the results are better when we use a pair of camera poses with enough baseline. In order to show the obtained results we follow the steps of our methodology working with a pair of images. We use images 1 and 5 with a baseline of 31.96 m:

- Extraction of features (SIFT). After using SIFT method in images 1 and 5 we get 136,232 keypoints and 94,775 keypoints respectively. As we explain in a previous section these keypoints are invariant to scale, pose, and changes of ilumination.
- Matching of points. The next step is the comparison of the two groups of keypoints. In this case we set a threshold of 0.6 for the distance between descriptors and we obtain 17,542 matching locations between image 1 and image 5. Figure 6 represents points (blue crosses) which have a matched point in the other image. Each represented point in left image has an equivalent point in the right image and *vice versa*.
- Robust matching through epipolar geometry. We use RANSAC method to calculate matrix **F** with SIFT matching points. The outliers percentage is very low, in this case there are two outliers and 15,540 inliers and a reprojection error average of 0.0594 pixels and a maximum of 3.36 pixels. In this case RANSAC use is not decisive, but it could be important in other examples to get robust matching.

- Camera processing. We show in Table 1 the reprojection error (in the control points E1, E2, ..., E9) obtained computing the cameras (P1, P2, ..., P9) using bundle adjustment method. In Figure 7 we can observe the camera poses respect the rock glacier.

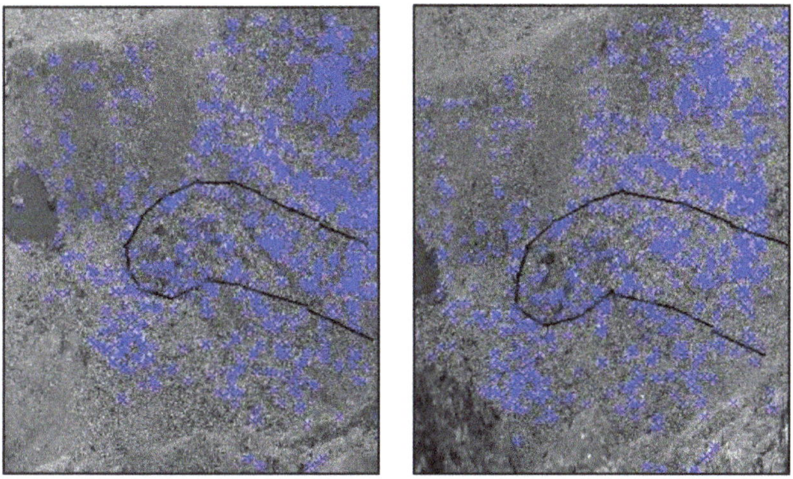

Figure 6: SIFT matched points denoted with crosses between two images of the Veleta rock glacier area.

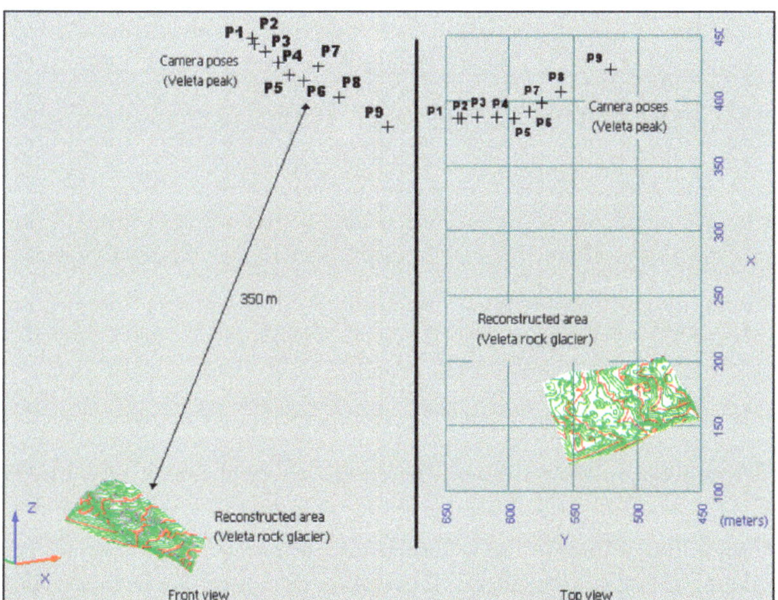

Figure 7: Camera pose estimation (front view and top view).

Table 1: Camera system accuracy in the year 2008

Cameras	Average (in pixels) respect points {E1, …, E9}
P1	0.69
P2	0.43
P3	0.53
P4	0.38
P5	0.53
P6	0.40
P7	0.45
P8	0.42
P9	0.63
Average all cameras (in pixels)	0.50

- DTM generation. Once we know the camera projection matrices we triangulate corresponding points to get a 3D reconstruction. We used the method propose in Section 4.5, getting a reprojection error average of 1.40 pixels and a maximun of 4.50 pixels. Therefore we have 15,540 points (X,Y,Z) of the rock glacier area which we can convert in a surface using a Delaunay triangulation (Figure 8).

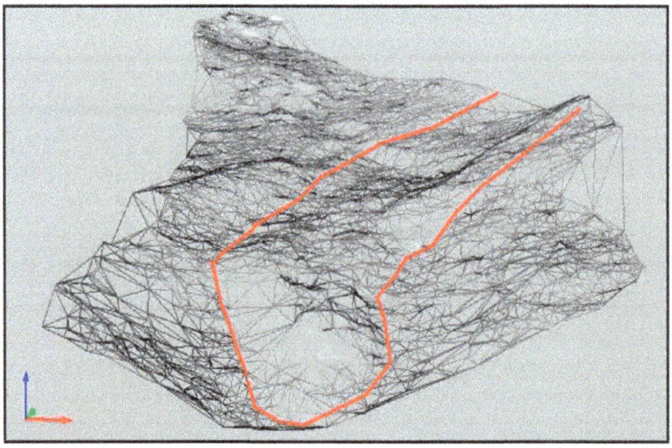

Figure 8: Example of Delaunay triangulation of the rock glacier (the view is oriented as in Figure 9).

We use cross sections of our reconstructed surface (15,540 points in a Delaunay triangulation surface) to calculate the obtained accuracy. We compare our cross sections with an accurate reconstruction (cross sections obtained using a total station Topcon GTS-502E). As we can see the results are acceptable (errors < 25 cm in height and planimetry) and therefore using this methodology we can get accurate surfaces

(Figure 9). We have developed necessary code using Matlab software for method implementation and we have used Photomodeler and Topcon Image Master for internal camera calibration.

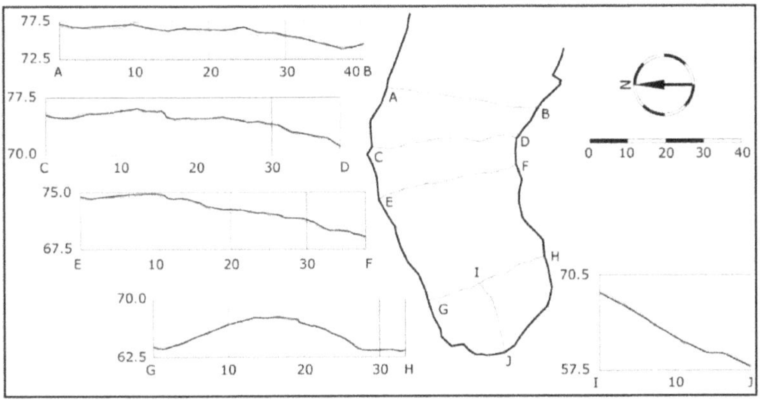

Figure 9: Comparison of our reconstructed surface cross sections (black polylines) and total station cross sections (blue polylines) and their situation in a Veleta rock glacier sketch.

CONCLUSIONS

We propose an automatic technique for rock glacier reconstruction using digital images. A significant amount of work in feature matching techniques has been reported in computer vision in recent years improving old techniques. From the results given by our methodology matching process we can conclude that the set of matches obtained is very good. Using this kind of techniques the post processing cost and time needed to analyze image information of geomorphologic structures can be dramatically reduced avoiding the arduous task of manual photogrammetric restitution.

Usually the images are taken with the same camera but the proposed methodology presents the advantage that it works the same using several types of camera, being each one defined by its own calibration. Thus, the cartography will be automatically produced even if the photographs are taken with different cameras or their external geometry includes variations in orientation or scale. Besides, the non-metric basis of the fundamental matrix used for the robust matching step makes possible to work with different and uncalibrated cameras for obtaining the final set of point matches.

Another advantage of the proposed method is that it deals with convergent images to produce a DTM of rock glaciers. It allows one to work with different types of cameras with acceptable results. The results quality also depends on the camera configuration in the acquisition phase, where large image base pairs with convergent views produce better results.

The presented method offers several advantages for the study of geomorphologic structures in terms of cost and time. Thus, we do not need flights of planes to obtain the photographs. The geodetic data are obtained in short time and we avoid the manual photogrammetric restitution in the office. Using this methodology points are automatically matched between images obtaining that nearly 100% of the matches are correct. Therefore, it is a robust method for the automatic 3D restitution to be used in the production of geomorphologic maps. In order to show the validity of the methodology experiments with different pairs of images of the rock glacier have been performed and the most representative are shown. Future work is focused toward dense matching and also on extending this method to images taken at different periods of time to study the evolution of the rock glacier. The objective is to have a cheap and fast tool for studying the evolution of the rock glacier along the time, which is our current research goal.

ACKNOWLEDGEMENTS

The authors are grateful for the collaboration on the geodetic data collection at the Veleta rock glacier to Antonio Gómez Ortiz, Ferran Salvador Franch (University of Barcelona) and Alan D.J. Atkinson Gordo (University of Extremadura). The authors wish to thank to the anonymous reviewers for their creative suggestions in this manuscript. The financial support for this research is from a Nacional Parks Project (Ministry of Natural Environment of Spain, Ref. 018/2007).

REFERENCES

1. Sanjosé, J.J.; Atkinson, A.D.J.; Salvador, F.; Gómez, A. Application of geomatic techniques to monitoring of the dynamics and to mapping of the Veleta rock glacier (Sierra Nevada, Spain). *Zeitschift für Geomorphologie* 2007, *51*, 79–89.
2. Serrano, E.; Sanjosé, J.J.; Agudo, C. Rock glacier dynamics in a marginal periglacial high mountain environment: Flow, movement (1991–2000) and structure of the Argualas rock glacier, the Pyrenees. *Geomorphology* 2006, *74*, 285–296.
3. Sanjosé, J.J.; Lerma, J.L. La fotogrametría digital: Una herramienta idónea para el cartografiado y modelado de zonas de alta montaña. In *Congreso de periglaciarismo en montaña*; Potes, Spain, 2001; pp. 185–203.
4. Corripio, J.G. Snow surface albedo estimation using terrestrial photography. *Int. J. Remote Sens.* 2004, *25*, 5705–5729.
5. Gómez, A.; Schulte, L.; Salvador, F.; Palacios, D.; Sanjosé, J.J.; Atkinson, A.D.J. Deglaciación reciente de Sierra Nevada. Repercusiones morfogenéticas, nuevos datos y perspectivas de estudio futuro. *Cuadernos de investigación geográfica de la Universidad de la Rioja* 2004, *30*, 147–168.
6. Bauer, A.; Paar, G.; Kaufmann, V. Terrestrial laser scanning for rock glacier monitoring. In Proceedings of the 8th International Conference on Permafrost, Zurich, Switzerland; 2003; pp. 55–60.
7. Kenyi, L.W.; Kaufmann, V. Estimation of rock glacier surface deformation using SAR interferometry data. *IEEE TGARS* 2003, *41*, 1512–1515.
8. Roer, I.; Kääb, A.; Dikau, R. Rockglacier acceleration in the Turtmann valley (Swiss Alps): Probable controls.*Norwegian J. Geography* 2005, *50*, 157–163.

9. Kääb, A.; Haeberli, W.; Gudmundsson, G.H. Analysing the creep of mountain permafrost using high precision aerial photogrammetry. 25 years of monitoring Gruben rock glacier. Swiss Alps. *Permafrost Periglac.* 1997, *8*, 409–426.

10. Kääb, A.; Kneisel, C. Permafrost creep within a recently deglaciated glacier forefield: Muragl, Swiss Alps. *Permafrost Periglac.* 2006, *17*, 79–85.

11. Kaufmann, V.; Ploesch, H. Mapping and visualization of the retreat of two cirque glaciers in the Austrian Hohe Tauern National Park. In XIX ISPRS Congress International Archives of Photogrammetry and Remote Sensing, Amsterdam, The Nederlands, July 16–22, 2000; 2000; Volume XXXIII, Part B4. pp. 446–453.

12. Kaufmann, V.; Ladstädter, R.; Kienast, G. 10 Years of monitoring of the Doesen rock glacier (Ankogel group, Austria)—A review of the research activities for the time period 1995–2005. In 5th ICA Mountain Cartography Workshop, Bohinj, Slovenia, March 30–April 1, 2006; pp. 129–144.

13. Sanjosé, J.J.; Lerma, J.L. Estimation of the rocks glaciers dynamics by environmental modeling and automatic photogrammetric technic. In XX Congress International Society for Photogrammetry and Remote Sensing, Commission VII, Istanbul, Turkey, July 12–23, 2004; Vol. XXXV-B7.

14. Gómez, A.; Palacios, D.; Ramos, M.; Casas, A.; Sanjosé, J.J.; Atkinson, A.D.J. Permafrost, evolución de formas asociadas y comportamiento térmico en el corral del Veleta (Sierra Nevada, España), Últimos resultados. *Boletín de la Real Sociedad Española de Historia Natural (Sección Geológica)* 2004, *99*, 47–63.

15. Mikhail, E.M.; Bethel, J.S.; McGlone, J.C. *Introduction to Modern Photogrammetry*; John Wiley: NY, USA, 2001.

16. Wolf, P.R.; Dewitt, B.A. *Elements of Photogrammetry with aplications in GIS*; McGraw Hill: Boston, MA, USA, 2000.

17. López-Nicolás, G.; Guerrero, J.J.; Sanjosé, J.J. Automatic photogrammetry for the production of geomorphologic cartography. In Sixth International Conference on Geomorphology, Zaragoza, Spain, September 7–11, 2005.

18. Lowe, D. Distinctive Image Features from Scale-Invariant Features. *Int. J. Comput. Vision* 2004, *60*, 91–110.

19. Zhang, Z.; Deriche, R.; Faugeras, O.; Luong, Q.T. A robust technique for matching two uncalibrated images through the recovery of the unknown epipolar geometry. *Artificial Intelligence* 1995, *78*, 87–119.

20. Guerrero, J.J.; Sagüés, C. Robust line matching and estimate of homographies simultaneously. In Iberian Conference on Pattern Recognition and Image Analysis, 2652 Lecture Notes in Computer Science, Mallorca, Spain, June 4–6, 2003; pp. 297–307.

21. Remodino, F. Detectors and descriptors for photogrammetric applications. In Photogrammetry and Computer Vision, Bonn, Germany, September 20–22, 2006.

22. Hartley, R.; Zisserman, A. *Multiple View Geometry in Computer Vision*, 2nd ed.; Cambridge University Press: NY, USA, 2004.

23. Rousseeuw, P.; Leroy, A. *Robust regression and outlier detection*; John Wiley: New York, NY, USA, 1987.

24. Fischler, M.A.; Bolles, R.C. Random sample consensus: A paradigm for model fitting with applications to image analysis and automated cartography. *Commun. Assn. Comput. Mach.* 1981, *24*, 381–395.

25. Triggs, B.; McLauchlan, P.; Hartley, R.; Fitzgibbon, A. Bundle adjustment – A modern synthesis. In Proceeding International ICCV Workshop on Vision Algorithms, Corfu, Greece, September 21–22, 1999; Springer: Berlin/Heidelberg, Germany, 2000; 1883, pp. 298–372

26. Torr, P.H.S. *A Structure and Motion Toolkit in Matlab*; Technical Report MSR-TR-2002-56; Microsoft Research: Cambridge, UK, 2002.

INDEX

A

aerial photogrammety 209
aerotriangulation 16, 18, 22
American Society of Photogrammetry and Remote Sensing (ASPRS) 132

C

commercial off the shelf (COTS) 16

D

Digital Elevation Models (DEMs) 5
Digital Terrain Model (DTM) 208

E

epipolar constraint 211, 212, 213, 214
Equilibrium Lines of Altitude (ELA) 4

F

Federal Emergency Management Agency (FEMA) 129, 132
Flight Controller Board (FCB) 82

G

Glacier Lake Outburst Floods (GLOF) 1, 4
Glaciers 1, 2, 3, 4, 6, 7, 8, 9, 10
ground control points (GCPs) 16, 18, 28, 32
Ground Control Points (GCPs) 5
Ground Sample Distance (GSD) 132

M

Multi Scale Curvature Classification (MCC) 27

N

Near-Infrared (NIR) 133
Normalized Differenced Water Index (NDWI) 4

O

Original Equipment Manufacturers (OEMs) 83

R

Remotely operated Piloted Air System (RPAS) 82
root mean square error (RMSE) 5

S

Scale Invariant Feature Transform (SIFT) 22
Southern Patagonia Icefield (SPI) 2
Southwest Florida Water Management District (SWFWMD) 129

T

Tie Points (TPs) 5

U

unmanned aircraft system (UAS) 20, 33